ENDLESS PERFORMANCE

BUILDINGS FOR PERFORMING ARTS

观演建筑

殷倩 编 常文心 译

辽宁科学技术出版社

Endless Performance

Buildings for performing arts at different scales and styles provide places for people where they can gather and appreciate performances of music, dance, drama and so on. These buildings for performing arts are the endless performances that directed by architects, who are also the playwright and endow the buildings with diversity. These buildings, as actors, become the best landmark that represent a city's culture features through rich colours, different appearances of day and night, and spaces feeling like flowing crotchets.

Whatever new or renovated buildings, or building for school, shopping centre, most beginning situations for architects are complex with many affective elements that need careful consideration, "and buildings for the performing arts are no exception. Indeed the geometry of the auditorium and performance area, the extent of services, the technical necessities, and the public expectations make these particular building types even more complex than the majority".

Regarding the layout, architects should fully consider "satisfy site, circulation, construction, environmental, statutory, economic and other requirements by the clients and local governments". General considerations affecting the building design include relationships between functions, external access, means of escape, sub-

永不落幕的演出

为表演艺术服务的建筑在规模与风格上多种多样，是为人们提供聚会和欣赏音乐、舞蹈、戏剧等表演艺术的场所。这类建筑，即我们在这里所说的观演建筑本身就是一幕幕永不落幕的演出。建筑师们作为演出的编剧和导演，赋予了建筑更多元化的表现方式，使它们以丰富色彩、日与夜的多姿造型、如音符般跃动的空间流动感成为最能代表城市文化气质的地标。

无论是哪一种建筑，新建的、改造的，或是学校、商场，建筑师在设计之初面临的各种影响设计方案的因素都不是简单的，而观演建筑中涉及的剧场（礼堂）与表演区域的几何学、服务功能的宽度和广度、音效和视觉效果所必须达到的技术支持以及公众的期待程度都使观演类建筑相较于其他绝大多数建筑类型更为复杂。

在建筑的布局设计方面，建筑师们需考虑满足地点、空间流动、环境、相关法规、经济以及来自客户和当地主管部门的其他要求和约束。通常来讲，影响观演类建筑设计的因素主要包括：各功能区之间的关系、遇险疏散的方式、内部功能细分、功能定位兼顾灵活性、声学设计策略、运行动力、消防、安全、通风、采暖、照明、通信、管道与排水、清洁与废物处

divisions, phasing and flexibility, acoustic strategy, energy strategy, fire protection, security, ventilation, heating, lighting, communications, plumbing and drainage, cleaning and refuse, plant rooms, integration and distribution of services, structure, internal finishes, windows, doors, fittings and equipment, signage and works of art, external works, form and approvals.

Agora Theatre in Netherlands, which was designed by UN Studio has typical Dutch features. "Both inside and outside walls were faceted, and all of the façade have sharp angles and jutting planes, which are covered by steel plates and glass, often layered, in shades of yellow and orange", just like those wonderful orange memories that were created by Dutch soccer team in the World Cup field. It's really very Dutch.

Due to special requirements of space and acoustics, most buildings for performing arts are designed following basic principles of traditional concert hall and opera house, but architects tried many new material and additional function strategies, and determined by energy conservation and sustainability considerations. Bernard Tschumi Architects applied

理、设备室、综合服务区与分散服务区、结构构造、内部修缮、窗体、门、相应配置与设施、标识与艺术品摆放、外部处理、形态以及如何获得方案审批通过。

UN Studio打造的荷兰阿格拉剧院有着明显的荷兰特色，墙壁的内外两侧都进行了切面处理，外墙尖锐的棱角和凸起的平面由黄色和橙色的钢板和玻璃板覆盖，一如世界杯赛场上荷兰军团留给人们的诸多橙色记忆，这是非常荷兰的建筑作品。

由于大剧院对空间、声学等方面有特殊要求，因此，大多数项目都遵循了传统音乐厅、大剧院的基本设计理念，但在材料以及附加功能上都进行了更多的尝试与创新，并且更多地考虑到节能和可持续性。伯纳德·曲米建筑师团队设计的法国里摩日音乐厅采用了全新的材料策略，外层表皮由弧形木条和透明的聚碳酸酯板制成，内层表皮由木材制成，而自然通风的设计使大厅能保持恒温，几乎不需要附加加热系统。艾皮特兹工作室设计的柯达伊中心除了包含音乐厅与排练室外，还有交响乐团的办公

new material strategies for the project of Limoges Concert Hall – "the outer envelope is made of wood arcs and translucent rigid polycarbonate sheets and inner envelope of wood." The natural ventilation that is integrated into keep the foyer at a temperate level, with little additional heating required. Építész Stúdió added office, café bookstore, lounge, etc. to Kodaly Centre except for concert hall, rehearsal room, and made the building become a mix-use leisure and art centre. When iPhone and iPad change our life, to appreciate elegant arts will not be limited in fixed and giant hall. Coop Himmelb(l)au created a transportable opera space, the Pavilion 21 Mini Opera Space in Munich, Germany. It is dismountable, transportable and re-mountable with distinctive shape, and moreover, realises the interior spatial acoustics, bringing music and elegant arts everywhere.

Thanks these respectable architects for bringing us these endless performances, which are played by buildings and architectural arts.

室、会议中心，咖啡厅、书店、休息室，使其成为一个方位的休闲艺术中心。在iPhone，iPad改变生活方式的时代中，对高雅音乐与表演艺术的欣赏也可以不被局限在坚固、硕大的建筑空间里，库伯·西梅布芬在德国慕尼黑打造的21号迷你歌剧院就是一个可移动的音乐景观，具有可拆卸性、可移动性和可重装性，造型抢眼，同时保证了室内音响效果，使高雅的艺术表演真正深入到每一个角落。

感谢可敬的建筑师们，以建筑物和建筑艺术为演员，为我们奉献了这些永不落幕的演出。

Contents
目 录

哈尔帕雷克雅维克音乐厅和会议中心

Harpa Reykjavik Concert Hall and Conference Centre

Location: Reykjavik, Iceland **Designer:** Henning Larsen Architects and Batteriid Architects **Completion date:** 2011 **Photos©:** Nic Lehoux **Gross floor area:** 28,000 square metres

项目地点：冰岛，雷克雅维克 设计师：亨宁·拉尔森建筑事务所和巴特利德建筑事务所 完成时间：2011年 摄影师：尼克·里奥科斯 面积：28,000平方米

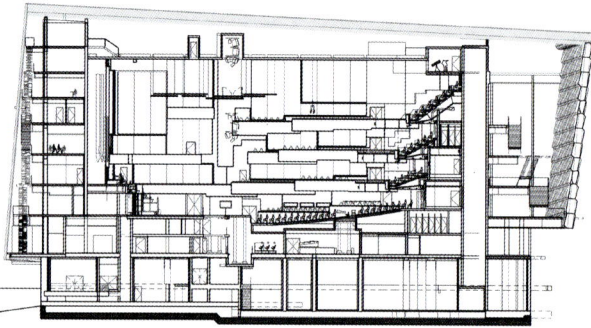

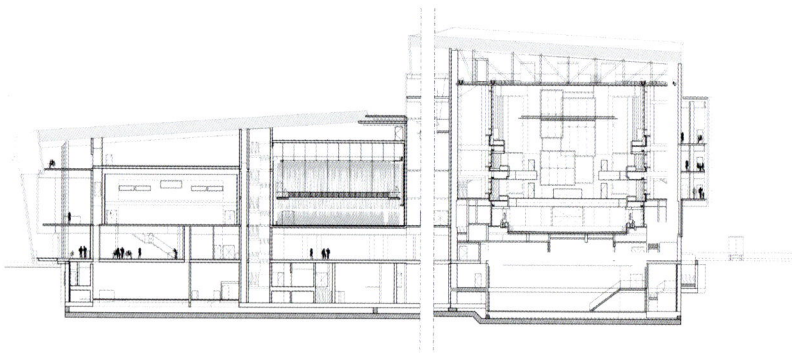

Situated on the border between land and sea, the Centre stands out as a large, radiant sculpture reflecting both sky and harbour space as well as the vibrant life of the city. The spectacular façades have been designed in close collaboration between Henning Larsen Architects, the Danish-Icelandic artist Olafur Eliasson and the engineering companies Rambøll and ArtEngineering GmbH from Germany.

The Concert Hall and Conference Centre of 28,000 square metres is situated in a solitary spot with a clear view of the enormous sea and the mountains surrounding Reykjavik. The Centre features an arrival and foyer area in the front of the building, four halls in the middle and a backstage area with offices, administration, rehearsal hall and changing room in the back of the building. The three large halls are placed next to each other with public access on the south side and backstage access from the north. The third floor is a multifunctional hall with room for more intimate shows and banquets.

Seen from the foyer, the halls form a mountain-like massif that similar to basalt rock on the coast forms a stark contrast to the expressive and open façade. At the core of the rock, the largest hall of the Centre, the main concert hall, reveals its interior as a red-hot centre of force.

项目坐落在陆地和海洋之间，巨大的放射形玻璃结构映衬着蓝天碧海和生机勃勃的城市生活。建筑宏伟壮丽的外立面由亨宁·拉尔森建筑事务所、丹麦裔冰岛艺术家奥拉夫尔·艾利亚森和德国工程公司朗姆波尔和艺术工程公司合作打造。

音乐厅和会议中心总面积28,000平方米，位于环绕雷克雅维克的辽阔海洋和山脉之间。中心由建筑前的到达和门廊区域、建筑中间的四个大厅和建筑后部的后台区域（包括办公室、行政空间、排练厅和更衣室）组成。其中三个大厅并列设置，公众从南侧进入，后台则从北侧进入。四楼是一个适用于更加私密演出和宴会的多功能大厅。

从门廊看去，大厅形成了山形模块，与海岸上的玄武岩类似，和具有表现力的开放式外立面形成了显著对比。建筑中央最大的主音乐厅暴露出它的室内设计，形成了红色的重心。

1. Overall view of the theatre   1. 剧院全景
2. Façade detail   2. 外立面细部

1

1. Front façade
2. Terrace
3. Entrance plaza and façade detail

1. 建筑正面
2. 平台
3. 入口广场和外立面细部

3

Ground level
1. Main entrance
2. Foyer
3. Information/tickets
4. Coat rack
5. Café/restaurant
6. 4th hall / Kaldalón
7. Flexible space – exhibition, dining etc.
8. Conference
9. Shop
10. Box office area
11. Loading dock
12. Back stage entrance
13. Technical space

一层平面图
1. 主入口
2. 门厅
3. 信息台/售票处
4. 衣帽间
5. 咖啡厅/餐厅
6. 第四大厅/卡尔达隆
7. 灵活空间——展览、就餐等
8. 会议室
9. 商店
10. 售票区
11. 装卸码头
12. 后台入口
13. 技术区

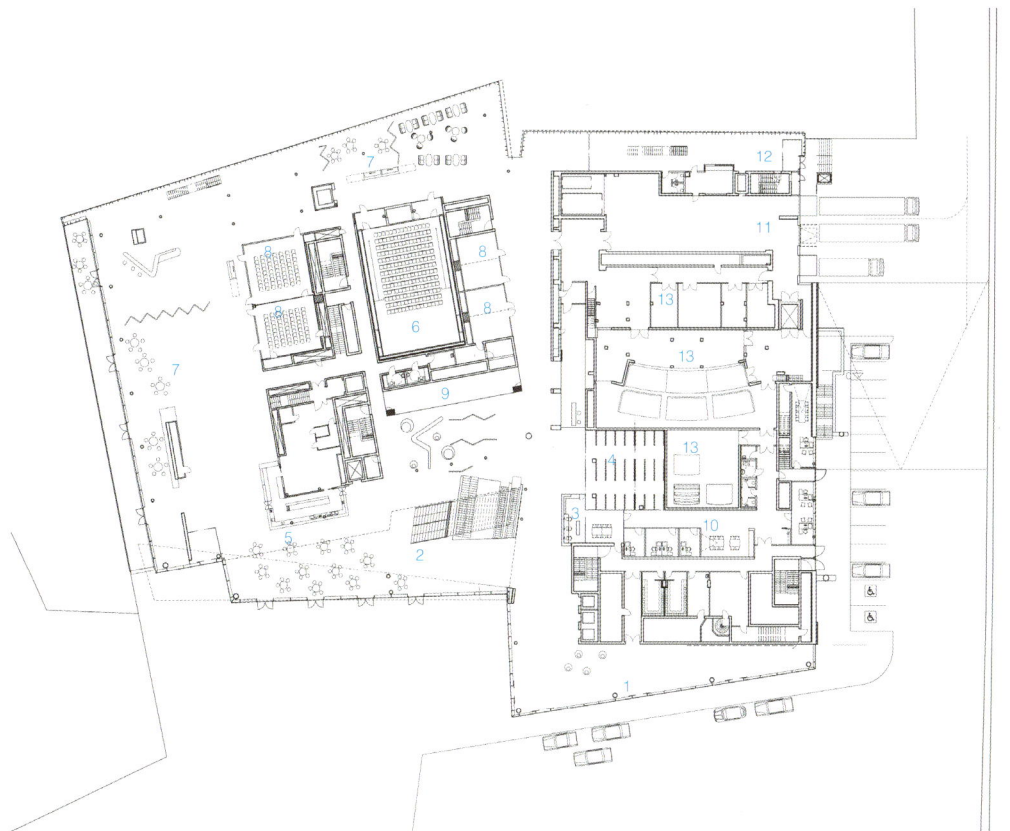

1. Ceiling and curtain wall detail
2. Lounge area
3. Interior detail

1. 天花板和幕墙细部
2. 休息区
3. 室内细部

1. Wall and ceiling detail
2. Lounge area
3. Stairs

1. 墙壁和天花板细部
2. 休息区
3. 楼梯

1. Seating detail
2. Theatre hall

1. 坐席细部
2. 剧院大厅

阿尔蒙特剧院 Almonte Theatre

Location: Huelva, Spain **Designer:** Donaire Arquitectos **Completed date:** 2010 **Photos©:** Fernando Alda
Gross area: 3,266 square metres

项目地点：西班牙，维尔瓦 设计师：多奈尔建筑事务所 完成时间：2010年 摄影师：费尔南多·阿尔达 面积：3,266 平方米

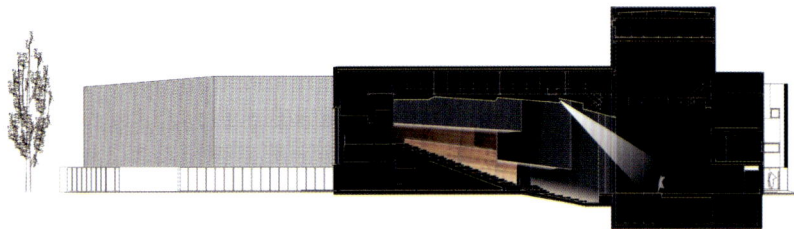

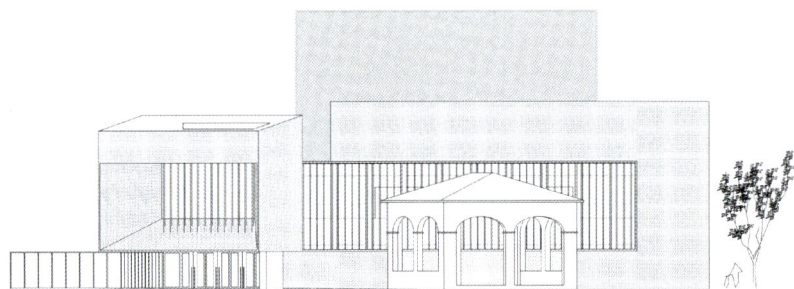

Located in Huelva, Spain, the Almonte Theatre by Donaire Arquitectos is on the site of an old winery. As a cultural institution, the building allows the architects to play with light and volume, which they've done with great success. It is a beautiful space for the arts.

The building is located on the site of an old winery. It has the challenge of integrating the existing old buildings, declared as cultural interest, and being part of a cultural complex of total three buildings and a common space. This space turns into the main place of the town and an important meeting area.

It is an opportunity to work on light, material and space. The path chosen to work on these concepts, is the contrast – contrast between outside and inside, between old and new, including a monumental scale and human scale, and the journey as the thread that sews and explains the intervention. A large area is covered with large proportions and controlled height works as a high threshold. A monumental scale lobby welcomes the visitor showing the scale of a public building.

多奈尔建筑事务所设计的阿尔蒙特剧院位于西班牙的维尔瓦，作为一座文化设施，项目为建筑师提供了巧妙运用光影和空间的机会。建筑师出色地完成了这个任务，打造了漂亮的艺术空间。

建筑场地原来坐落着一家旧酿酒厂。建筑师面临将原有建筑融合到由三座建筑和公共空间组成的城市综合体里。整个空间成为了城市的主要景点和重要的集会区域。

项目结合了光影、材料和空间设计。这些概念所采用的方法形成了对比——内外对比、新旧对比、宏伟规模和人性尺度的对比。整个旅程像一条线，引领着人们进行探索。一大片覆盖着大规模和受控高度的设计的区域形成了临界空间。巨大的大厅欢迎着来访者，展示了公共建筑的规模。

1. Front courtyard　1. 前庭
2. Main entrance　2. 主入口

1

2

1 Main entrance
2. Side view of the entrance
3. Entrance lobby

1. 主入口
2. 入口侧面
3. 入口大堂

Ground floor and level 1 function area
1. Ticket hall
2. Hall and exhibition
3. Auditorium
4. Coffee shop
5. Dressing
6. Toilets
7. Wardrobe

一层和二层功能区
1. 售票厅
2. 大厅和展览区
3. 礼堂
4. 咖啡店
5. 更衣室
6. 洗手间
7. 衣帽间

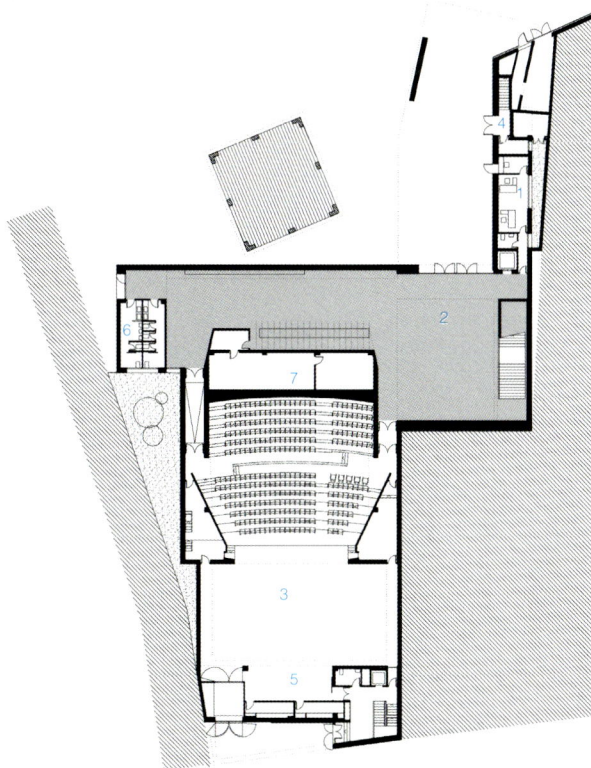

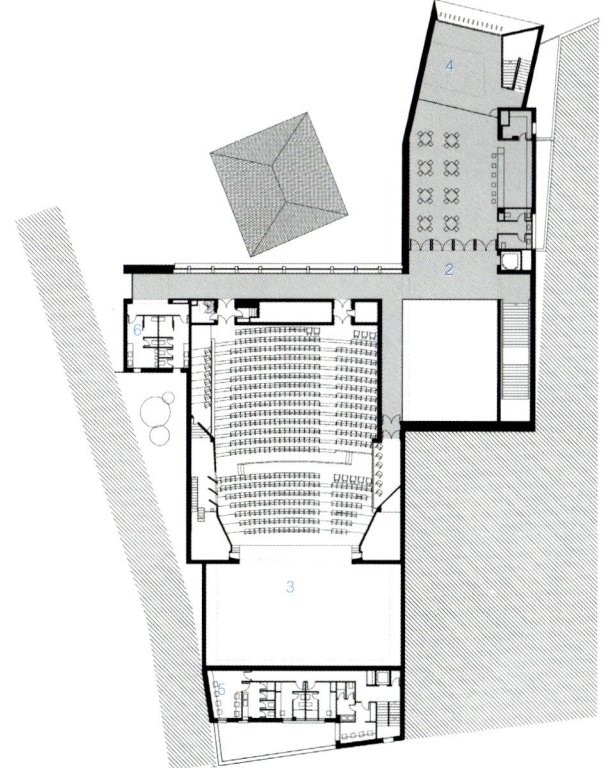

1. Reception
2. Reception and lounge area
3. Atrium

1. 前台
2. 前台和休息区
3. 中庭

2

3

2

1-3. Staircases

1-3. 楼梯

3

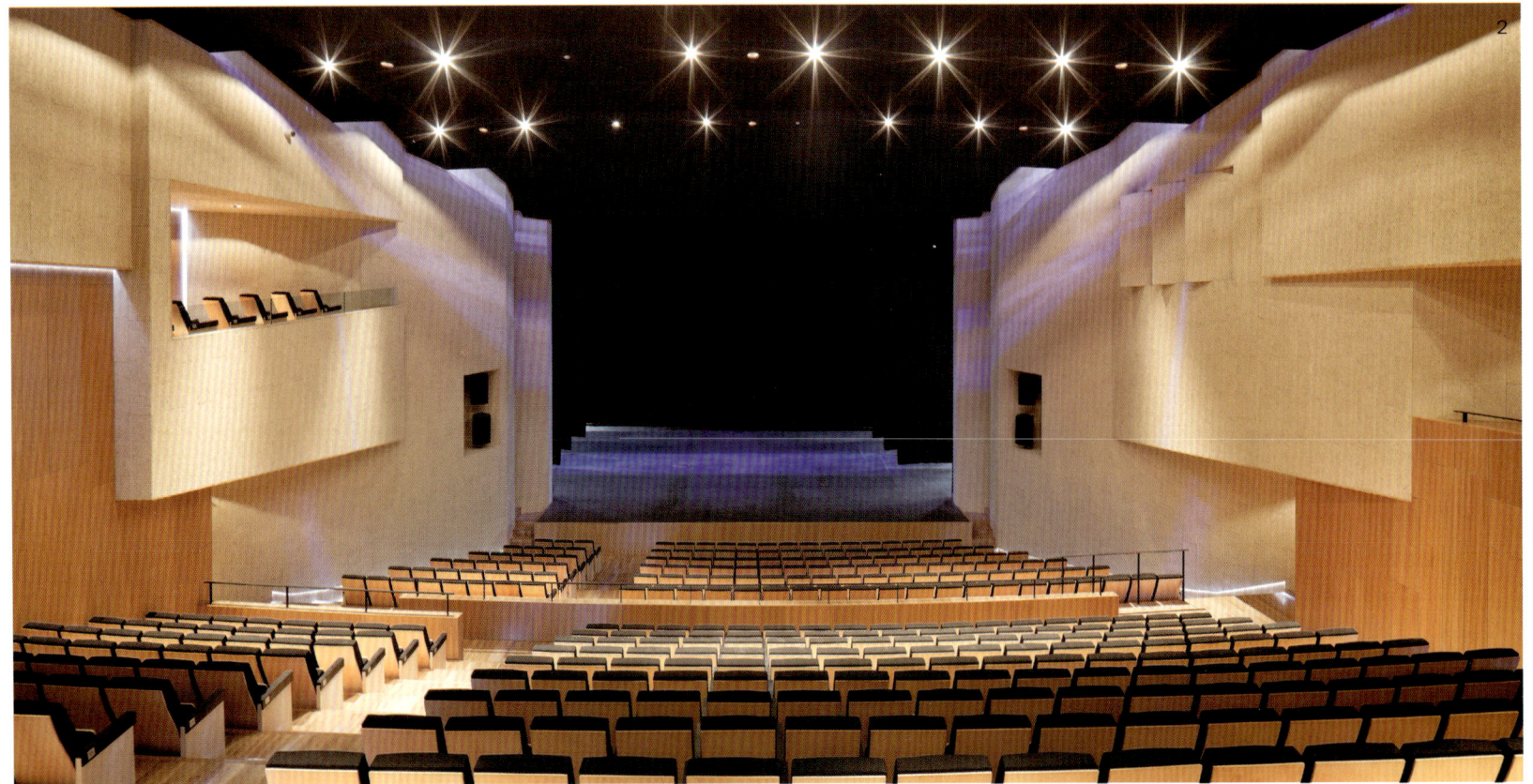

3

1. Auditorium viewed from the stage
2. Stage viewed from the seating
3. Seating detail

1. 从舞台看礼堂
2. 从观众席看舞台
3. 观众席细部

多尔勃－米斯塔西尼剧院 # Dolbeau-Mistassini Theatre

Location: Quebec, Canada **Designer:** Paul Laurendeau Architecte **Completion date:** 2008 **Photos©:** Marc Gibert/adecom.ca **Gross floor area:** 2,630 square metres

项目地点：加拿大，魁北克 设计师：保罗·劳伦德建筑事务所 完成时间：2008年 摄影师：马尔科·吉伯特/adecom.ca 面积：2,630平方米

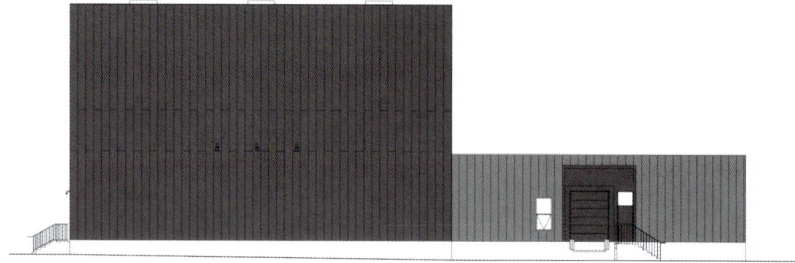

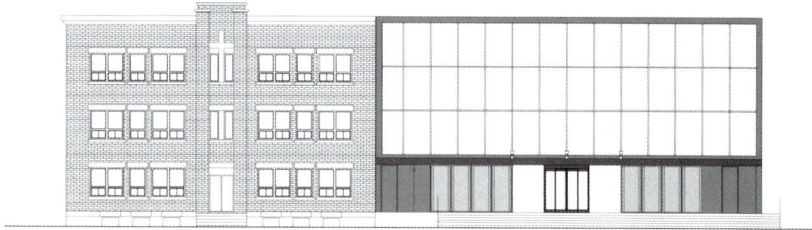

In 2006, Paul Laurendeau Architecte, in association with the office of Jodoin Lamarre Pratte et Associés Architectes, was declared winner of the national architectural competition to design the 491-seat Dolbeau-Mistassini Theatre.

The building is strategically located in the centre of town, assuming its function of major cultural institution. The plan is organised along a central axis that establishes a symmetrical layout and the major circulation path for the public, from the main entrance doors to the auditorium across the foyer. Its street façade uses elements found in theatre language: a canopy, a grid of exposed white lights, black paint, poster display cases and signage.

The foyer is dimensioned to host a variety of uses, from exhibitions, conferences, banquets to parties and special events. Soundproof vestibules isolate the auditorium from adjoining spaces. Based on a cylindrical form, the auditorium features a gently sloping orchestra of 11 rows and 293 seats topped by a flat acoustic ceiling with a huge crystal chandelier. Two semi-circular balconies of 2 rows and 99 seats increase the height effect and give drama when spectators arrive. Red, black and gold colours strengthen the image of the performance space.

The building was opened in 2008. Population, artists and press greeted it with great enthusiasm. Season tickets were sold in 15 minutes.

2006年，保罗·劳伦德建筑事务所与朱德因·拉马里·普莱特建筑事务所合作，获得了拥有多尔勃－米斯塔西尼剧院（可容纳491人）国家建筑竞赛的优胜。

建筑坐落在城市中心，凸显了本身作为文化建筑的地位。项目沿着中轴规划，形成了对称式布局，从主入口穿过门厅到礼堂，为公众提供了主流通路径。建筑的临街面采用了传统剧院设计元素：天篷、裸露的白灯、黑色漆面、海报展示柜和引导标示。

门厅具有许多功能，从展览、会议、宴会到派对和特别活动。隔音门廊将礼堂与邻近区域隔开。礼堂以圆柱造型为基础，以11排乐队演奏处和293个坐席为特色，二者上方的隔音天花板上装饰着巨大的水晶吊灯。两个半圆形双排包厢和99个坐席增添了空间的高度感，为观众营造了戏剧效果。红色、黑色和金色突出了表演空间的形象。

建筑于2008年开放。大众、艺术家和媒体对它呈现了巨大的热情。剧院的季度票在15分钟内被抢购一空。

1. View from the yard   1. 庭院
2. View from the parking area   2. 停车场
3. Side view at night   3. 夜晚建筑侧面
4. Entrance at night   4. 入口夜景

3

4

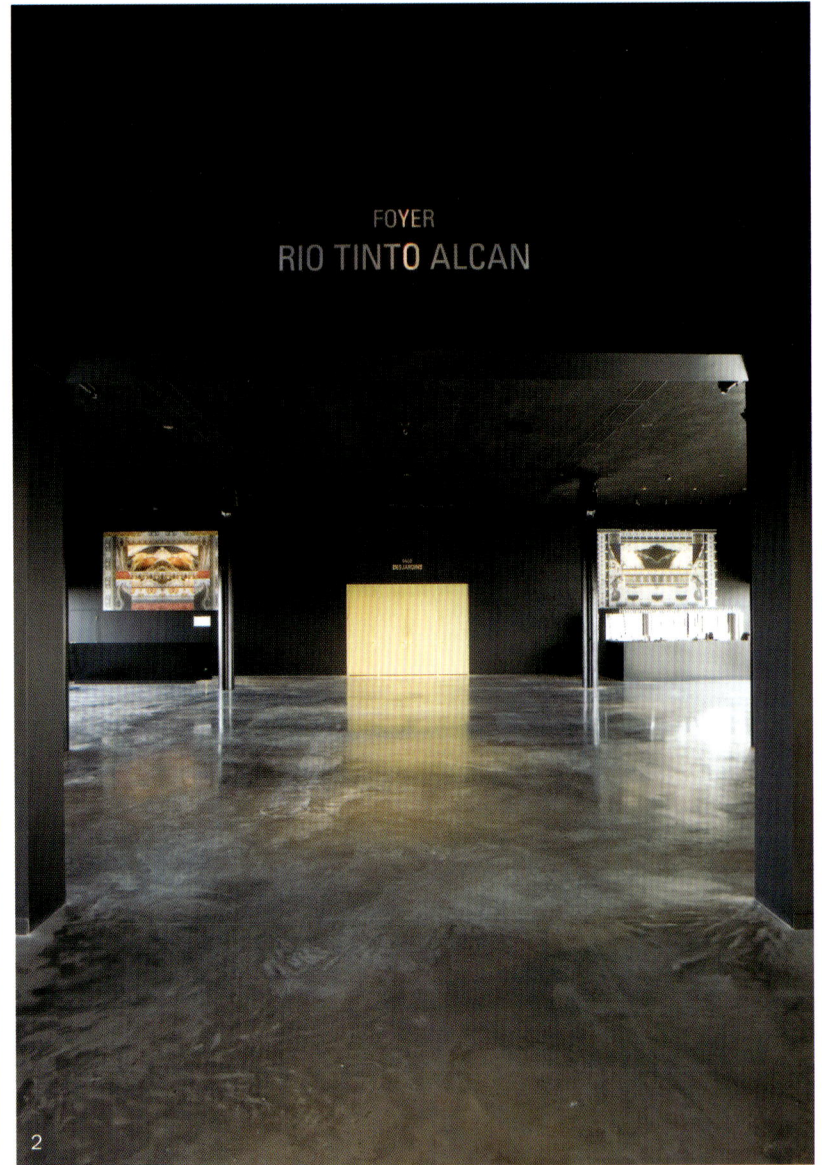

1. Entrance
2. Entrance lobby
3. Waiting area/lounge

1. 入口
2. 入口大堂
3. 等候区/休息区

Ground floor plan
1. Hall
2. Porter
3-6. WC Men
7. WC Women
8. OMB room
9. Stair
10. Sprinklers

一层平面图
1. 大厅
2. 门房
3-6. 男士洗手间
7. 女士洗手间
8. 管理办公室
9. 楼梯
10. 洒水装置

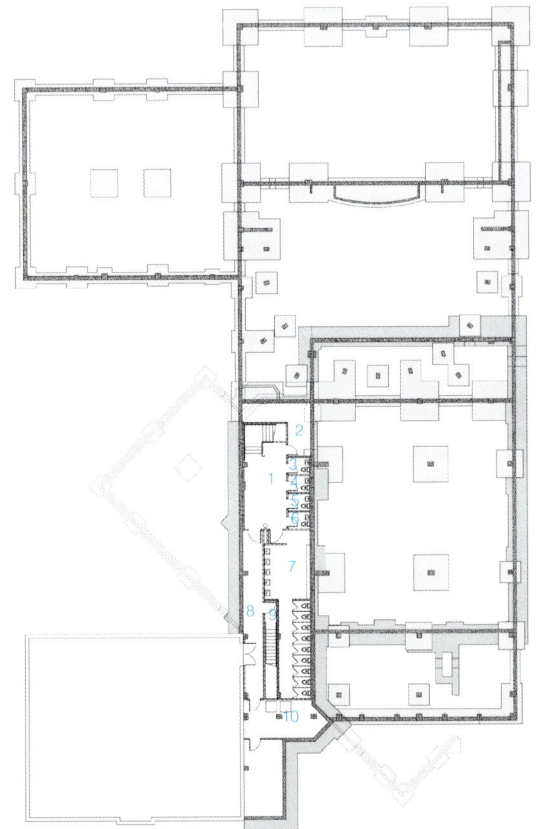

SALON VERT
AbitibiBowater

1 Detail of auditorium ceiling and seating
2. The stage
3. Auditorium viewed from the upper level

1. 礼堂天花板和坐席细部
2. 舞台
3. 从楼上看礼堂

实验媒体和表演艺术中心

Experimental Media and Performing Arts Centre

Location: City of Troy, USA **Designer:** Grimshaw Architects **Completion date:** 2008 **Photos©:** Aaron Esto; Paul Rivera

项目地点：美国，特洛伊城 设计师：格雷姆肖建筑事务所 完成时间：2008年 摄影师：亚伦·埃斯托；保罗·里维拉

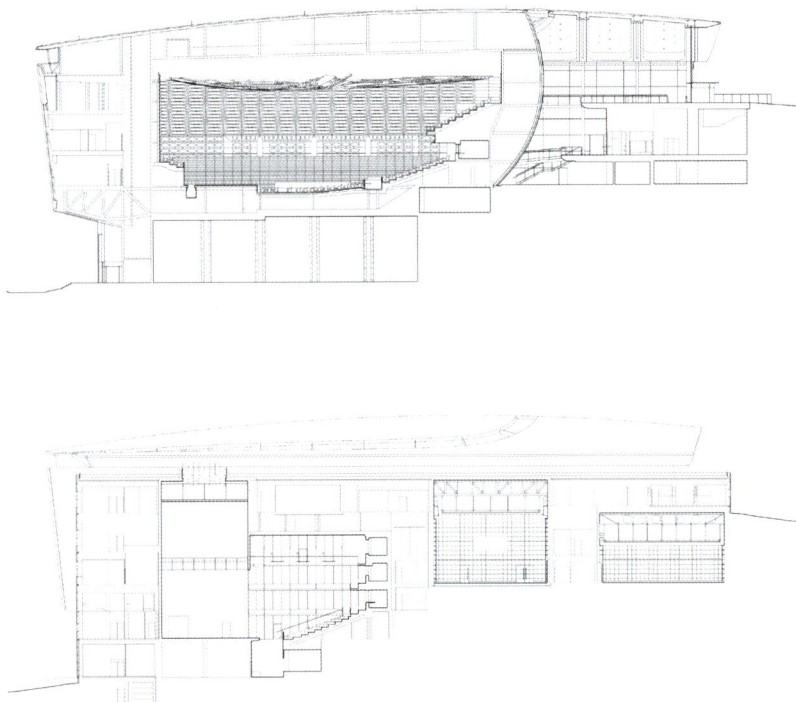

The EMPAC (Experimental Media and Performing Arts Centre) programme poses a question – how to combine, in one building, the permanence of the traditional performing arts with the necessarily transient character of experimental media. As one of their starting points, Grimshaw considered the resonant chambers of stringed instruments, in the belief that tradition and experimentation are linked by the unvarying physics of sound. So that the traditional and the experimental may be seen as yoked together yet distinct, Grimshaw arranged the concert hall and atrium axially with the main entrance in a linear sequence on the north side of the building, while the studios and theatre form an adjacent sequence on the south. A conceptual dialogue was then initiated between these two sequences by seeing the concert hall manifested as the physical presence of an object in space, while the theatre and studios represent the physical absence of discovered voids within a solid.

By taking advantage of the slope of the hillside site, the design solves one of the persistent challenges of performing arts projects: concealing the windowless mass of a very large hall and fly tower.

Because the main entrance is at hilltop level, close to the roof, while the volume of the concert hall is fitted into the slope below, a large "found space" opens up between the two. Upon entering the building, visitors find themselves at the top of the atrium and main circulation area, looking down at the exterior of the concert hall.

Access to the concert hall is provided via elevated walkways that span like gangplanks across the atrium. The entire hull of the concert hall is contained within the atrium, allowing public circulation all around it.

The entire north façade of the building is a glass curtain wall, providing transparency between the EMPAC interior and the city of Troy. The glass wall allows daylight to flood the atrium, augmented by a halo skylight around the top of the concert hall that washes the cedar hull with the changing light of the day. By night, the wood hull is lit up within the building and creates an iconic external identity that can be seen from distance.

实验媒体和表演艺术中心提出了一个问题——如何在同一座建筑中结合传统表演艺术和实验媒体。建筑师考虑了弦乐器的共鸣室，认为传统和实验可以由永恒的乐声连接，二者结合在一起，却又相互独立。建筑师将音乐厅和中庭与主入口设在同一个轴线上，位于建筑的北侧，而工作室和剧院则在南侧排列。这两排空间之间形成了对话，音乐厅体现了空间内的实体，而剧院和工作室则呈现了实体内的空洞。设计利用山坡，解决了表演艺术项目中最重要的问题之一：将没有窗户的建筑实体（包括巨大的大厅和舞台塔）隐藏起来。

建筑的主入口设在山顶，靠近屋顶，音乐厅则设在下面的山坡上，二者中间设有一个大型空间。走进建筑，人们便置身于中庭和主流通区域的顶部，俯瞰着音乐厅的外部。

人们通过横跨中庭的跳板式高架走道进入音乐厅。音乐厅的外壳设在中庭内，让公共走道环绕着它。

建筑的整个北立面都采用玻璃幕墙，在表演艺术中心室内和特洛伊城之间形成了通透感。玻璃幕墙让阳光洒入中庭，而环绕着音乐厅屋顶的光环天窗增强了这一效果。夜晚，木质外壳从建筑内部被点亮，让建筑在远处也具有高度辨识度。

1. Front view at night　1. 夜间前景
2. Façade detail　2. 外立面细部

1

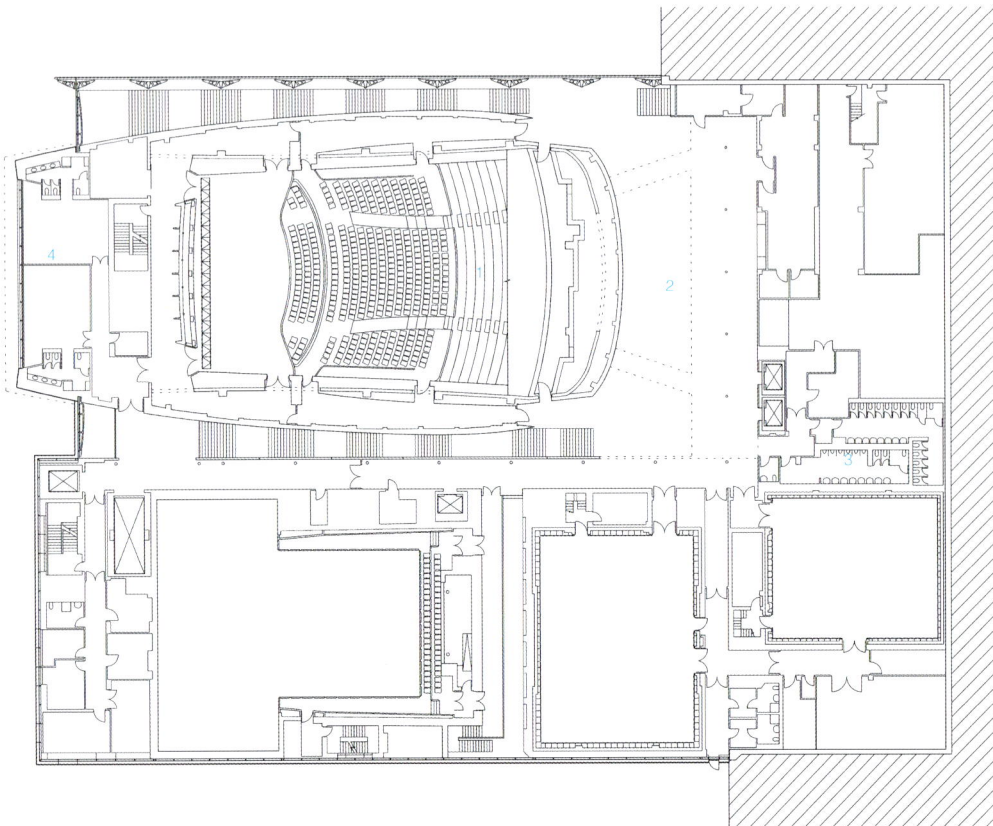

1 Hall
2 Lobby/lounge
3 Toilets
4 Dressing room

1. 大厅
2. 大堂/休息室
3. 洗手间
4. 更衣室

1. Interior viewed through the glass curtain wall
2. Upper level seating entrance
3. Theatre hall entrance detail

1. 透过玻璃幕墙看室内
2. 上层坐席入口
3. 剧院大厅入口细部

1. Seating detail
2. Café and lounge area
3. Rehearsal room

1. 坐席细部
2. 咖啡厅和休息区
3. 排练室

皇家音乐学校——泰勒斯表演学习中心

Location: Toronto, Canada **Designer:** Kuwabara Payne McKenna Blumberg Architects **Completion date:** 2009 **Photos©:** Eduard Hueber, Tom Arban **Area:** 17,651 square metres

项目地点：加拿大，多伦多 设计师：KPMB建筑事务所 完成时间：2009年 摄影师：爱德华·赫伯尔；汤姆·阿本 面积：17,651平方米

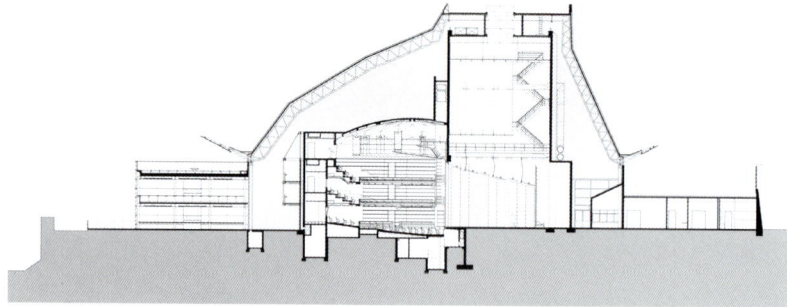

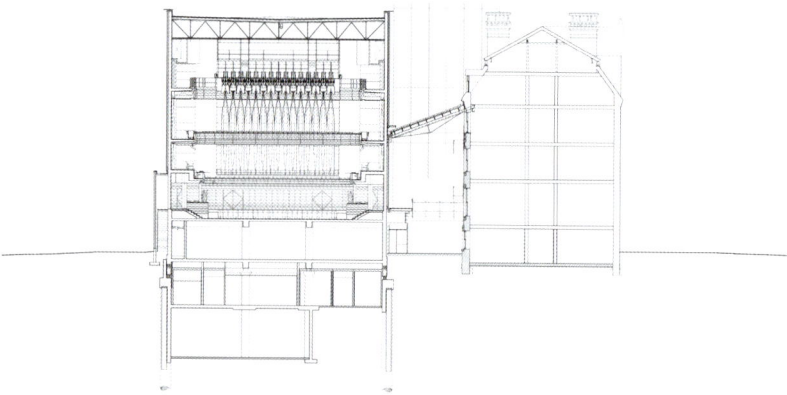

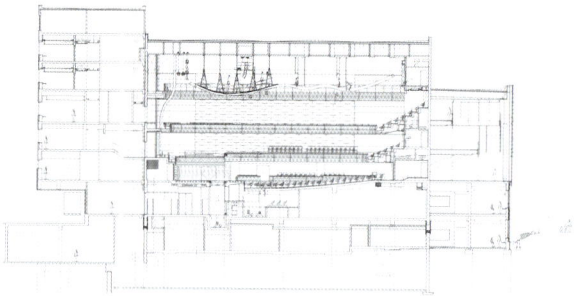

The Royal Conservatory, TELUS Centre for Performance and Learning

The overall project involved the progressive restoration of McMaster Hall and the construction of a new TELUS Centre for Performance and Learning to create a unique hybrid of a teaching and rehearsal facility and destination concert venue with three major performance venues. The space between the historic and new building is enclosed to create a skylit pedestrian court linking the Bloor Street entrance to the Concert Hall and Lobby. The glass and steel structure of the new addition provides a dynamic counterpoint to the polychromatic façades of the heritage buildings.

Urbanistically, the project occupies an important site in mid-town Toronto at the threshold of the University of Toronto's downtown campus and integrates Philosopher's Walk, a landscape pedestrian route that runs north and south linking Bloor Street to Hoskin Avenue. The design was strategically conceived to define a new cultural precinct for the City in concert with the transformation of the adjacent Royal Conservatory and the expansion of the Gardiner Museum around the corner on Queen's Park.

Although the new additions are substantive in scale and size, the siting, massing and articulation is deferential to the 19th century heritage buildings on Bloor Street, which have housed the RCM since 1962. The emphasis on transparency and contemporary building systems create a dynamic counterpoint to the polychromatic masonry walls when encountered from Philosopher's Walk. KPMB was also involved in the restoration of the exterior heritage fabric and 240-seat Ettore Mazzoleni Hall.

A key mandate was to maximise the capacity and flexibility for integrating new technology and adapting to changes and growth in programmes. The new additions include 43 new teaching and practice studios, the renovation of Ihnatowycz Hall (1898) and a new 150-seat Conservatory Theatre, a rehearsal space designed to accommodate a range of functions, from special events to the Learning Through the Arts. In scale and proportion it replicates the acoustic quality and stage size of the main Koerner Concert Hall to prepare students for live performance.

In addition to Mazzoleni Hall and Conservatory Theatre, the project incorporates the 1,135-seat Michael and Sonja Koerner Concert Hall. Koerner Hall is the performance heart of the project, and will provide a premiere acoustical environment. Its undulating wood "veil" integrated with the canopy above the stage define an iconic image for the RCM.

整个项目包含麦克马斯特大厅的翻新和新泰勒斯表演学习中心的建造工程，打造了一个独特的教学和排练设施和卓越的音乐表演空间。新旧建筑之间的空间被打造成一个封闭的步行长廊，连接了布鲁尔大街入口和音乐厅、大堂。新设施的玻璃钢铁结构与历史建筑多姿多彩的外立面形成了鲜明的对比。

项目占据了多伦多市中心一处重要的场地，位于多伦多大学市中心园区的入口，结合了哲人走道——一条从南北向连接布鲁尔街和霍思金大道的景观步行街。项目与邻近的皇家音乐学院改造工程和王后公园转角的加尼尔博物馆扩建工程一起，为城市打造了一个全新的文化街区。

尽管扩建工程在规模和尺寸上独立存在，其选址、概念设计和接合处都与19世纪的历史建筑相一致。布鲁尔街上的历史建筑从1962年起就一直是皇家音乐学院的所在地。设计的通透度和现代建筑系统与多彩的石造墙壁形成了鲜明的对比。KPMB建筑事务所同时还负责了历史建筑外墙和可容纳240人的埃多勒·马佐莱尼大厅的修复工作。

项目的主要要求之一便是打造能够采用新技术和适应未来发展变化的能力和灵活性。新建筑包含43间全新的教学联系工作室、伊奏多瓦兹大厅的改造（建于1898年）和可容纳150人的音乐学校剧院、一个多功能排练空间（可进行特殊活动和艺术表演学习）。建筑在规模和比例上复制了科尔纳尔音乐厅的音效品质和舞台尺寸，为学生们提供了现场表演准备训练。

除了马佐莱尼大厅和音乐学校剧院之外，项目还包含可容纳1,135人的迈克尔和索尼娅·科尔纳尔音乐厅。科尔纳尔厅是项目的表演核心，将提供非凡的音效环境。音乐厅的木罩与舞台顶部的天篷相结合，形成了皇家音乐学院的标志形象。

1. Surrounding street view   1. 周边街景
2. Access to the theatre   2. 剧院入口
3. General view along the street   3. 街面全景
4. Overall view from the opposite of the street   4. 从对街看建筑全景

1. Entrance lobby/reception/waiting area
2. Top view of the café

1. 入口大厅/前台/等候区
2. 俯瞰咖啡厅

Ground floor plan
1. Music court
2. Box office/entrance
3. Ihnatowycz Hall
4. Galleria
5. Café/bar
6. Children's programme
7. Philosopher's walk court
8. Studios
9. Lounge
10. Library
11. Orchestra lift
12. Back of house

一层平面图
1. 音乐庭院
2. 售票处/入口
3. 伊拿多瓦兹大厅
4. 连廊
5. 咖啡馆/酒吧
6. 儿童项目
7. 哲学家庭院
8. 工作室
9. 休息室
10. 图书馆
11. 乐队电梯
12. 后台

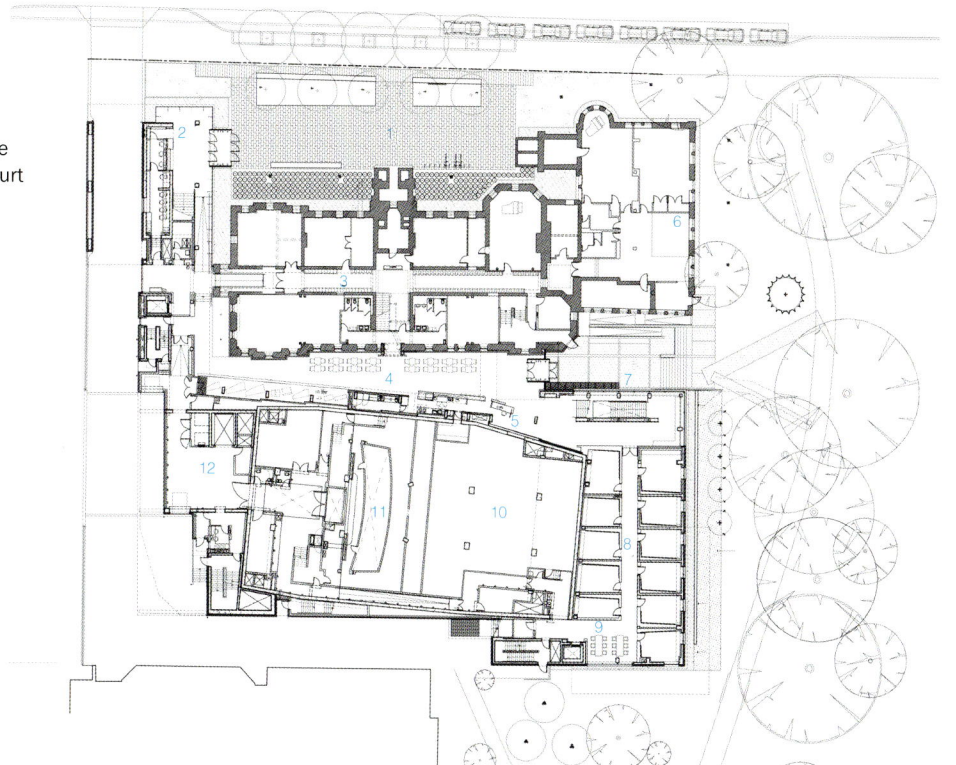

1. The café
2. Rehearsal room
3. Meeting room

1. 咖啡厅
2. 排练室
3. 会议室

1. Overview of the hall
2. Stage and ceiling detail
3. Seating detail

1. 大厅全景
2. 舞台和天花板细部
3. 坐席细部

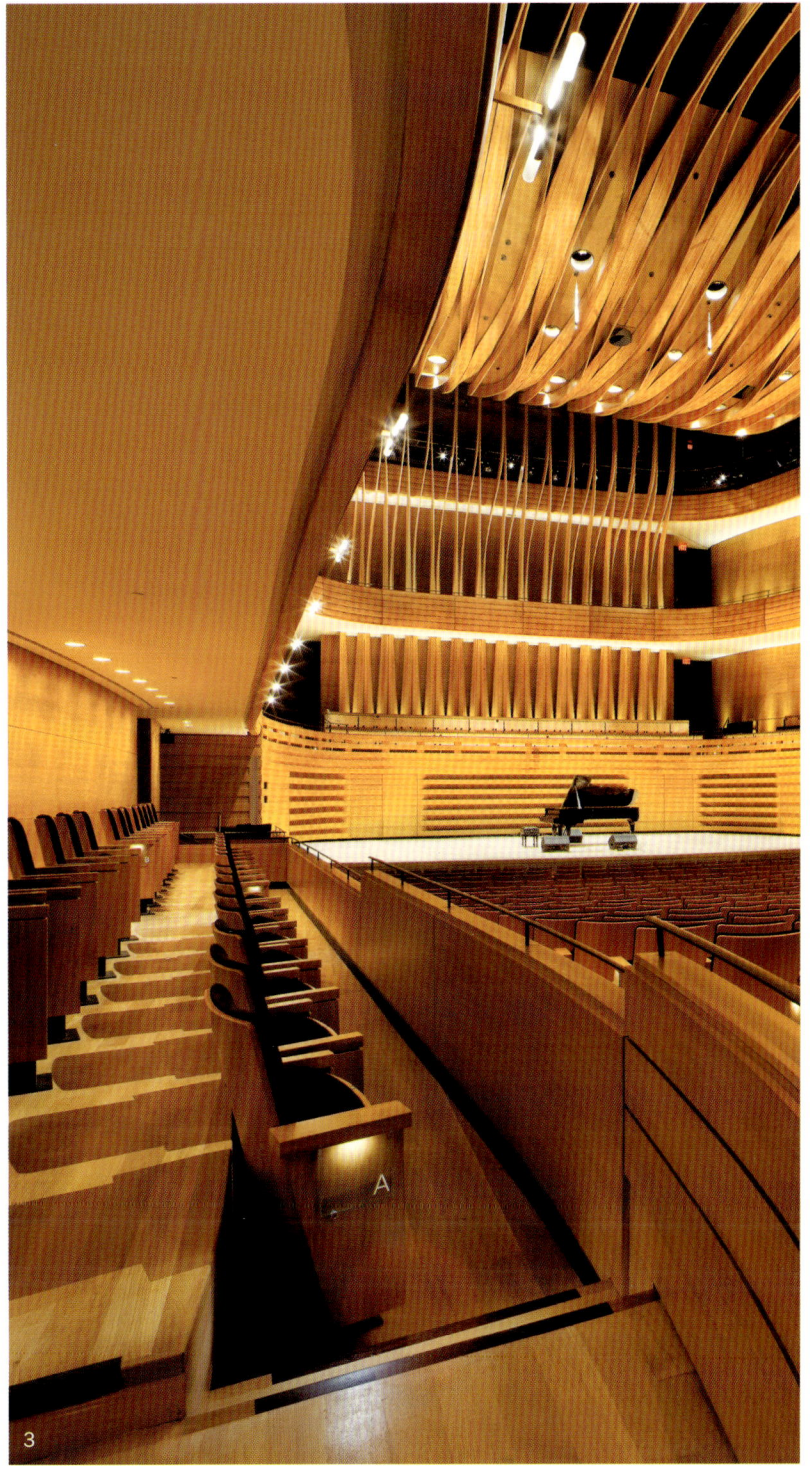

坦佩艺术中心 Tempe Centre for the Arts

Location: Tempe, USA **Designer:** ARCHITEKTON + Barton Myers Associates, Inc. **Completion date:** 2007 **Photos©:** Peter Robertson, Ohn Linden, Richard Krull, Michael Masengarb **Area:** 103,195 square metres **Awards:** National Council of Structural Engineers Association Excellence in Structural; Engineering Award – New Building $30-$100 Million, 2008; USITT United States Institute for Theatre Technology Merit Award, 2008; Valley Forward Environmental Excellence Crescordia Award, 2008; Valley Forward Environmental Excellence Design Award of Merit, 2008

项目地点：美国，坦佩 设计师：阿奇泰克顿+巴尔顿·迈尔斯建筑事务所 完成时间：2007年 摄影师：彼得·罗伯森；欧恩·林登；理查德·克鲁尔；迈克尔·马森加伯 面积：103,195平方米 所获奖项：2008年美国国家结构工程师委员会结构工程优秀奖——3,000万到1亿美元新建筑；2008年美国剧院技术协会优秀奖；2008年山谷先锋环境优秀奖克莱斯科尔蒂亚奖；2008年山谷先锋环境优秀奖荣誉设计奖

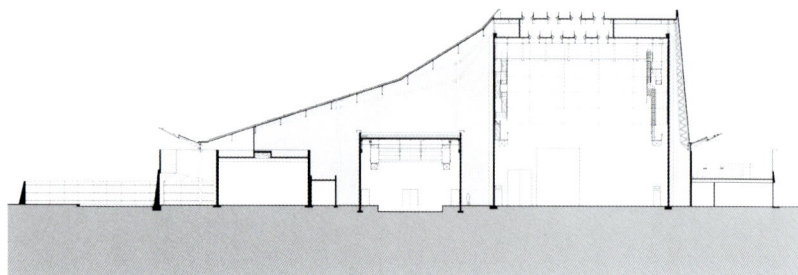

Acoustic mitigation through design was a critical concern because the site is directly below the flight path to Phoenix's Airport. Inspired by the jagged buttes of Monument Valley, the iconic, protective 16-pitch, 10-layer roof conceals the fly tower, provides acoustic mitigation and modulates natural light sheltering its patrons from the harsh desert sun. The roof also facilitates the collection of rainwater, channelling it into a stone pond on the north side, similar to the way arroyos deliver rivers of rain across the desert floor.

Designed as a collection of pavilions within a sculptural, protective envelope, the spacious lobby is an interior "town square" offering protection from the harsh environment and is like a theatre itself.

The building draws its formal inspiration from the native ruins at Pueblo Bonita at Chaco Canyon, which consisted of traditional clustering of circular rooms or "kivas". A unifying 91-centimetre thick circular concrete wall creates a 360-degree iconic landform honouring the traditions of indigenous tribes: the Anasazi and Chocóan practices of utilising circular floor plans within protective walls that surround a central plaza; and the Hohokam tradition of linking landscape to building with meandering paths.

Juxtaposing concrete with warm woods, tribal vocabulary is referenced through traditional metals, stone, colours and indigenous patterns throughout.

由于项目场地就设在凤凰城机场的飞行路线下方，隔音设计至关重要。屋顶受到纪念山谷起伏的孤峰的启发，分为10个层次的16个斜屋顶隐藏了舞台塔，提供了隔音效果，缓和了刺目的沙漠阳光。屋顶还便于收集雨水，将雨水引入北侧的一个石头池塘里，与小溪穿越沙漠将雨水导入河流相似。

宽敞的大堂被设计成一系列场馆的结合体，形成了一个室内广场，让人们免受外面严酷的环境侵害，就像剧院本身一样。

建筑造型灵感来源于查科峡谷的印第安村庄，由一系列传统的圆形空间（又名"大地穴"）组成。统一的91厘米厚的圆形混凝土墙壁打造了360度标志性地势，纪念了传统的土著部落。建筑师在环绕中心广场的防护墙内采用了环形楼面布局。霍霍坎文化传统通过漫步走道连接了景观和建筑。

部落设计将混凝土和温和的木材结合在一起，采用了传统金属、石材、色彩和本土图案。

1. North façade
2. Northeast view from Tempe Town Lake
3. West entrance on Tempe Town Lake walking path

1. 建筑北立面
2. 从坦佩城湖看建筑东北面
3. 坦佩城湖走道上的西入口

1

1. West entrance as seen from the Arts Park
2. Lobby with individual arts pavilions

1. 从艺术公园看西入口
2. 带有独立艺术展厅的大堂

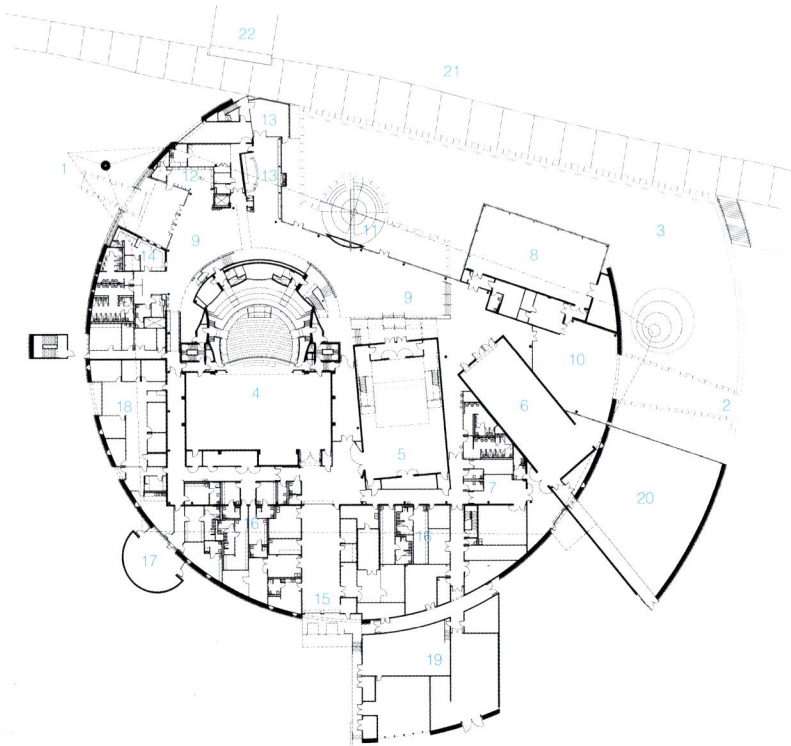

Ground floor plan	一层平面图
1. Main entrance	1. 主入口
2. Park entrance	2. 公园入口
3. Reflecting pool	3. 倒影池
4. 600-seat theatre	4. 600人剧院
5. 200-seat theatre	5. 200人剧院
6. Art gallery	6. 画廊
7. Gallery support	7. 画廊辅助设施
8. Multipurpose room	8. 多功能室
9. Lobby	9. 大堂
10. East entrance court	10. 东入口庭院
11. Outdoor fireplace & terrace	11. 露天壁炉和平台
12. Gift shop	12. 礼品店
13. Bar & outdoor café	13. 酒吧和露天咖啡馆
14. Box office	14. 售票处
15. Loading corridor	15. 装货走廊
16. Dressing rooms	16. 更衣室
17. Performers' courtyard	17. 表演庭院
18. Administrative offices	18. 行政办公室
19. Mechanical	19. 机械室
20. Outdoor sculpture garden	20. 露天雕塑公园
21. Tempe Town Lake	21. 坦佩城湖
22. Inflatable bridge	22. 充气桥

LAKESIDE

GALLE

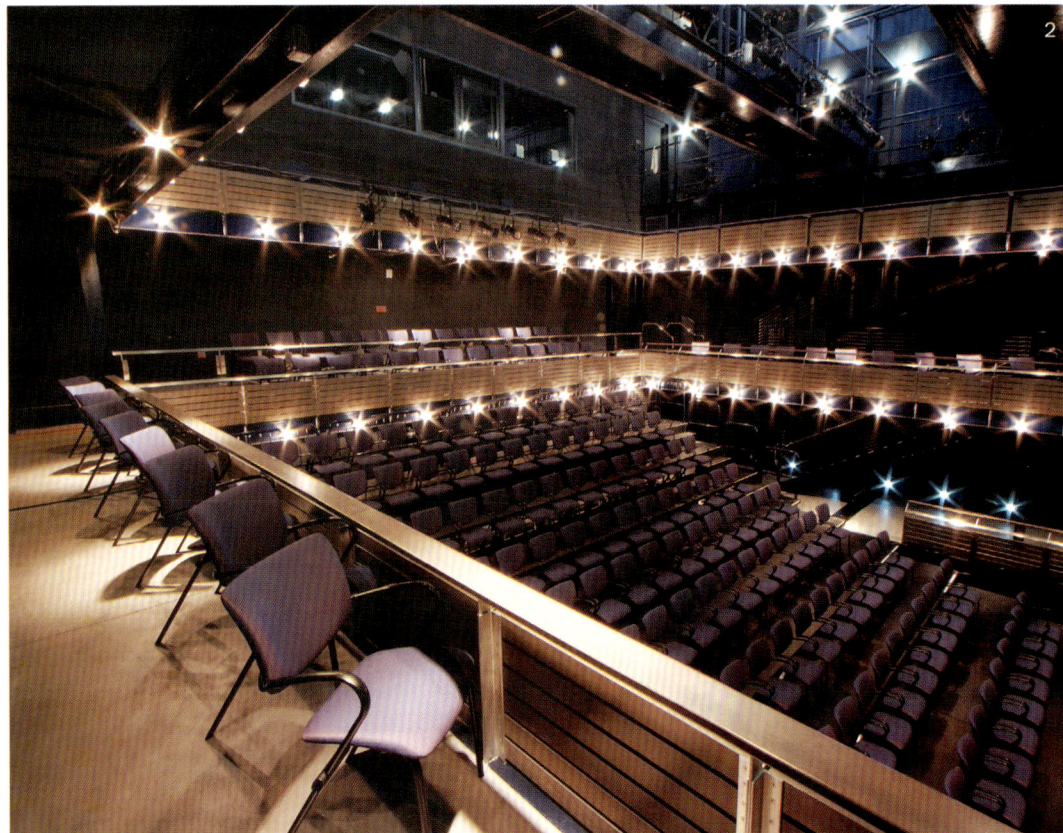

1. 600-seat main theatre
2. 200-seat flexible theatre
3. Art gallery

1. 600坐席主剧院
2. 200坐席灵活剧院
3. 艺术画廊

3

迪和查尔斯·威利剧院 # The Dee and Charles Wyly Theatre

Location: Dallas, USA **Designer:** REX/OMA, Joshua Prince-Ramus (partner in charge) and Rem Koolhaas, in collaboration with Kendall/Heaton Associates **Completion date:** 2009 **Photos©:** Iwan Baan, Tim Hursley

项目地点：美国，达拉斯 设计师：REX/OMA、约书亚·普林斯·拉姆斯（主管合伙人）和雷姆·库哈斯，Kendall/Heaton事务所合作设计 完成时间：2009年 摄影师：伊万·巴恩、蒂姆·赫斯利

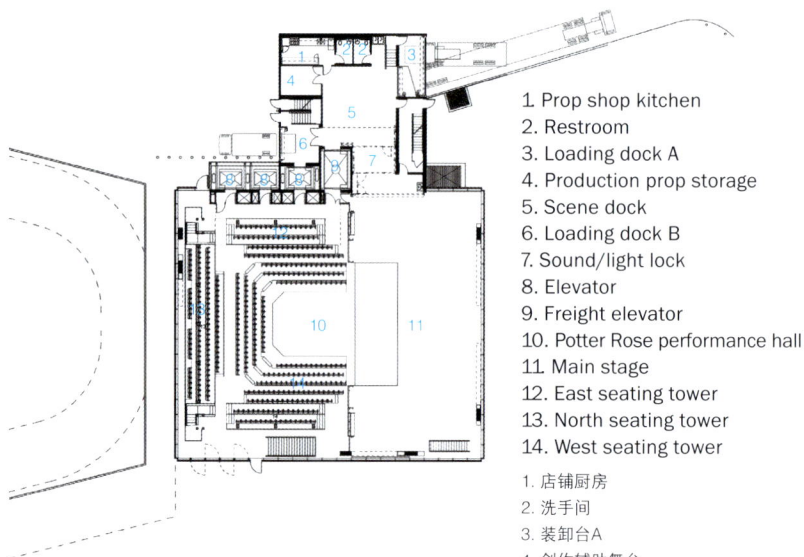

The Dee and Charles Wyly Theatre is one of four venues that comprise the AT&T Performing Arts Centre, a new performing arts centre for music, opera, theatre and dance. The 575-seat Potter Rose Performance Hall in the Wyly Theatre provides a new state-of-the-art home for the Dallas Theatre Centre and Dallas Black Dance Theatre, as well as numerous other performing arts organisations that serve Dallas and the region.

The Wyly Theatre is one of the world's most innovative theatre facilities. The 12-storey building features an unprecedented "stacked" design – a vertically organised facility that completely rethinks the traditional arrangement of a theatre's parts.

Unlike a typical theatre building, where support spaces wrap around the stage house, this unique design positions transitional, technical and work zones either above or below the auditorium, creating a highly flexible performance space. The facility's advanced mechanised "superfly" system can raise and lower both scenery and seating, allowing artistic directors to easily change the venue's configuration to best serve their artistic visions, choosing between proscenium, thrust and flat floor set-ups. The flexibility of the facility will allow the Wyly Theatre to host a wide range of classical and experimental drama, dance and musical productions, and world-renowned vocalists and dance troupes.

The Wyly Theatre is situated on the south side of the AT&T PAC's new Elaine D. and Charles A. Sammons Park, which embraces and unifies the four venues, creating a dynamic cultural destination in downtown Dallas. The exterior walls of the Potter Rose Performance Hall are made of an acoustic quality transparent glass curtain wall system with integral shade controls, which can be configured to create a virtually "seamless" vista of the outdoors, as well as to allow pedestrian views into the working operations of the theatre environment. The upper portions of the building are clad in pre-fabricated aluminum panels with random repetitions of vertical tubular extrusions, producing a textured effect.

The Wyly Theatre also includes a cocktail bar, rehearsal spaces, administrative offices, a costume shop, lobby, auditorium, stage support areas, mechanical rooms, production spaces and a rooftop multipurpose space.

1. Prop shop kitchen
2. Restroom
3. Loading dock A
4. Production prop storage
5. Scene dock
6. Loading dock B
7. Sound/light lock
8. Elevator
9. Freight elevator
10. Potter Rose performance hall
11. Main stage
12. East seating tower
13. North seating tower
14. West seating tower

1. 店铺厨房
2. 洗手间
3. 装卸台A
4. 创作辅助舞台
5. 布景存放处
6. 装卸台B
7. 声音/灯光控制
8. 电梯
9. 载货电梯
10. 波特·罗丝表演厅
11. 主舞台
12. 东侧坐席
13. 北侧坐席
14. 西侧坐席

迪和查尔斯·威利剧院是四个AT&T表演艺术中心之一，集音乐、歌剧、戏剧和舞蹈表演于一身。拥有575个坐席的波特·罗丝表演厅为达拉斯剧院中心、达拉斯黑人舞蹈剧院以及许多当地的表演艺术组织提供了最先进的设备。

威利剧院是世界上最具创新精神的剧院设施之一。12层的大楼以空前的"堆叠式"设计为特色——剧院的垂直组织结构完全颠覆了传统的剧院布局。

典型的剧院建筑将辅助空间设在舞台的四周，与此不同，威利剧院将过渡、技术和工作区设在礼堂的上下，打造了高度灵活的表演空间。剧院先进的机械化系统可以任意调整布景和座椅的高度，让艺术指导们方便地改变空间的布局，以适应他们的艺术视角，在舞台前部、中部和平地设施中进行选择。设施的灵活性让威利剧院可以上演各种经典和试验性的戏剧、舞蹈和音乐作品，迎接世界知名歌手和舞蹈团。

威利剧院位于AT&T新建的伊莱恩·D和查尔斯·A·萨门斯公园的南侧，公园环绕着这四个表演设施，为达拉斯市中心打造了一个动感的文化景点。波特·罗丝表演厅的外墙采用了透明隔音玻璃幕墙系统，结合了整体阴影控制，可以让室内与室外景观融为一体，也让行人可以直视剧院内部的环境。建筑的上层采用了预制铝板，散乱的垂直管状凸出营造出纹理效果。

威利剧院还包括鸡尾酒吧、排练空间、行政办公室、戏装商店、大厅、礼堂、舞台辅助区、机械室、创作室和屋顶多功能空间。

1. A overlooking of the theatre and its surroundings
2. The exterior of the theatre

1. 俯看剧院与周围建筑
2. 剧院外观

1

1. The dynamic design makes the theatre stands out of its surroundings
2. View of the interior from the street
3. Entrance hall
4. View from the Mark and Barbara Thomas Lemmon Rooftop Terrace

1. 极富动感的设计使剧院从周围建筑中脱颖而出
2. 从外部看剧院内部
3. 剧院入口大厅
4. 从马克和芭芭拉·托马斯·雷蒙屋顶平台向外看

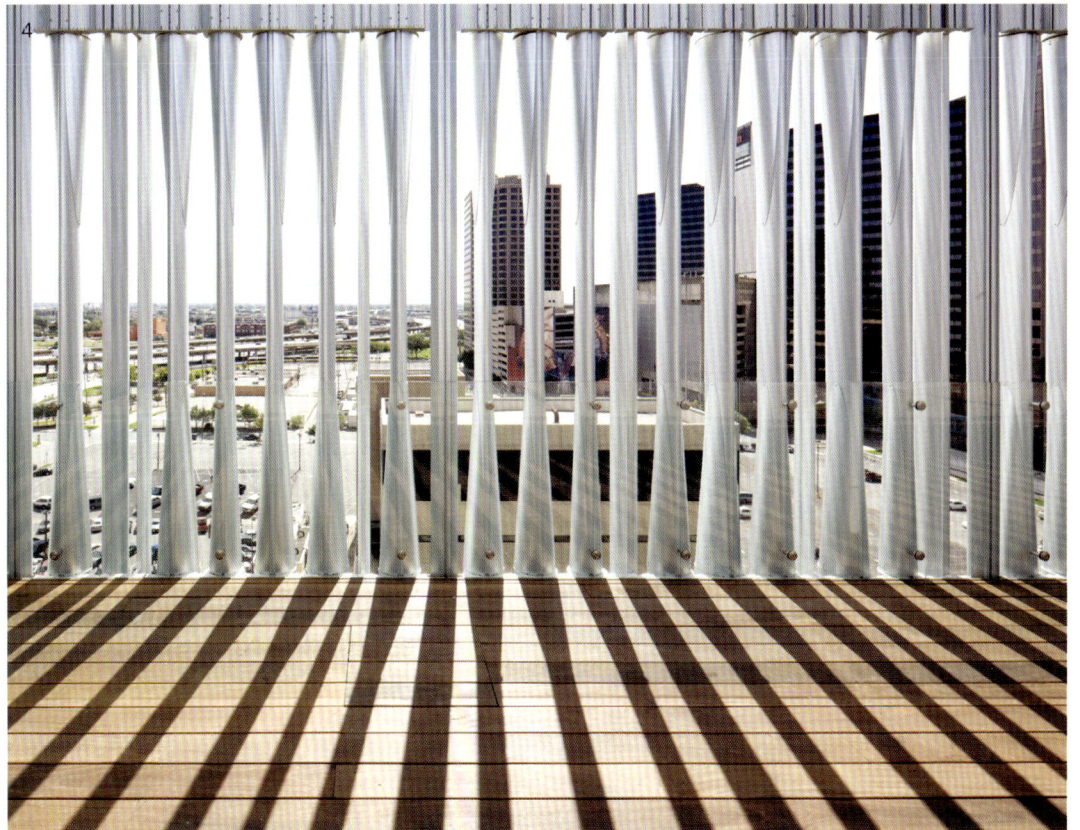

1. The Ann Swisher and Michael F. McGehee Education Centre overlooks the costume shop
2. The conference room doubles as the control room for the studio theatre
3. Stair of the performance hall

1. 安·史伟莎和迈克尔·F·麦克吉希教育中心俯瞰剧院里的戏服店
2. 会议室还可用作小剧场控制室
3. 表演大厅的楼梯

3

1

2

1. The patron lounge
2. The studio theatre
3. A stairwell in the performance hall
4. Conference room or control room

1. 赞助人休息室
2. 小剧院
3. 表演厅的楼梯
4. 会议室/控制室

1

2

1. Audience seats in the performance hall
2. The potter rose performance hall
3. The mechanical room
4. View from the upper level outdoor terrace

1. 表演厅的观众席
2. 波特·罗丝表演厅
3. 机械室
4. 从上层露天平台向外看

重庆大剧院 # Chongqing Grand Theatre

Location: Chongqing, China **Designer:** gmp – von Gerkan, Marg and Partners Architects **Completion date:** 2009 **Photos©:** Hans Georg Esch, Heiner Leiska (model)

项目地点：中国，重庆 设计师：GMP建筑事务所 完成时间：2009年 摄影师：海纳尔·利斯卡、莫德尔保·维尔纳、GMP、汉斯·乔治·埃斯科、本·麦克米伦

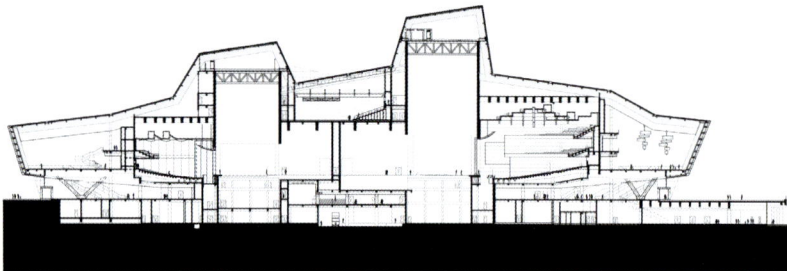

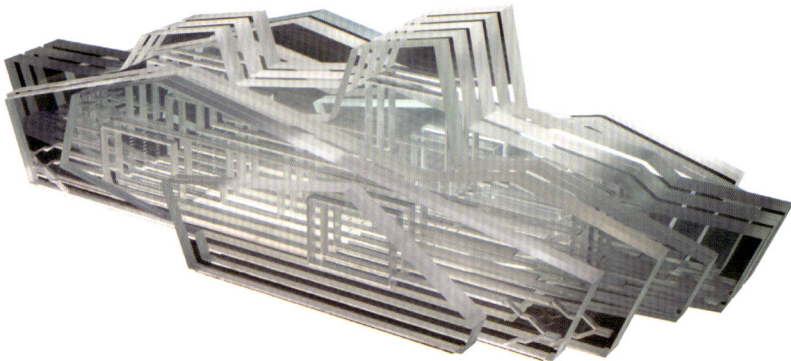

Theatre performances, musical dramas or operas are among the highest levels of artistic interpretation of the reality of human existence, and of dreams and illusions, wishes and pleasures. Going to the theatre means leaving the ordinary world behind. The performance can be at once pretentious and sophisticated, sometimes sublime and beautiful, occasionally cool, matter-of-fact or funny – but it is always a very special social occasion.

The architectural shell is intended to give structural expression to this extraordinariness and the world of illusion. An expressive sculptural array of parallel double-walled, spaced glass strips that jostle with and nudge up to each other generates in overview and side view a metaphorical image of a ship – an almost dramatic image of a theatre building sailing in a sea of light.

The Grand Theatre is close enough to the Yangtze River to seem to "float" on it. Surreal light reflections and light patterns create poetic compositions of reality and fiction comparable to the world of illusion of a theatre performance. The Grand Theatre lies majestically on a headland above the Yangtze opposite the peninsula with the city centre of the metropolis of Chongqing. The uniqueness of the location and the panoramic view of the imposing skyline of the city in conjunction with the sculptural architecture make this theatre a symbol, a place of international cultural encounters.

Concertgoers and theatregoers can reach the building from all directions via different transport modes. From the river, a ferry terminal on the extension to the grand boulevard provides transport. On the south side, there are the two entrance roads into the car parks for private cars. Thence, an access area with lifts and steps leads directly to the upper main foyer. The large plaza around the theatre is principally reserved for pedestrians. Only in the southern section is there a pick-up/set-down area for VIPs and taxis. The Grand Theatre Chongqing thus has no back. All sides and façades of the building are open to visitors and allow access to the theatre.

戏剧表演、音乐剧或歌剧是人类现实最高层次的艺术形式，是梦想和幻想、愿望和愉悦的体现。来到剧院边等同于离开了平凡的世界。表演可以做作而世故，也可以崇高而美好，或冷静或滑稽，但是它始终是一个特殊的社交场合。

剧院的建筑表壳要为它的特殊和奇妙提供结构表现。极具表现力的雕塑感双层玻璃墙排列结构结合起来，形成了轮船的造型，远远看去，宛如剧院在光海中航行。

大剧院临近长江，仿佛漂浮其上。超现实感的光反射和光带打造了如诗如画的效果，堪与戏剧表演的梦幻世界比肩。大剧院庄严地坐落在长江的河岬之上，遥望重庆市中心所在的半岛。独一无二的位置和壮观的城市全景让剧院成为了一个标志，是国际文化相遇的焦点。

观众们可以从各个方向、通过各种交通形式来到剧院。水上有林荫道延长线上的渡轮码头，陆上有两条私家车通道。电梯和台阶直通剧院的大厅。环绕着剧院的广场主要为行人保留。广场南侧是贵宾和出租车入口。重庆大剧院没有背面，四面和建筑的外墙都向游客开放。

1. The Grand Theatre lies on a headland opposite the peninsula with the city centre of the metropolis of Chongqing
2. The front of the theatre
3. The expressive sculpture architecture brings up a metaphor of ship

1. 江畔的大剧院，遥望重庆市中心所在的半岛
2. 大剧院正面
3. 富有表现力的雕塑式建筑造就了一个船体的隐喻

1. The hall of the theatre
2-4. The stairs connecting all levels of the theatre

1. 大剧院大厅
2~4. 大剧院内连接各层的楼梯

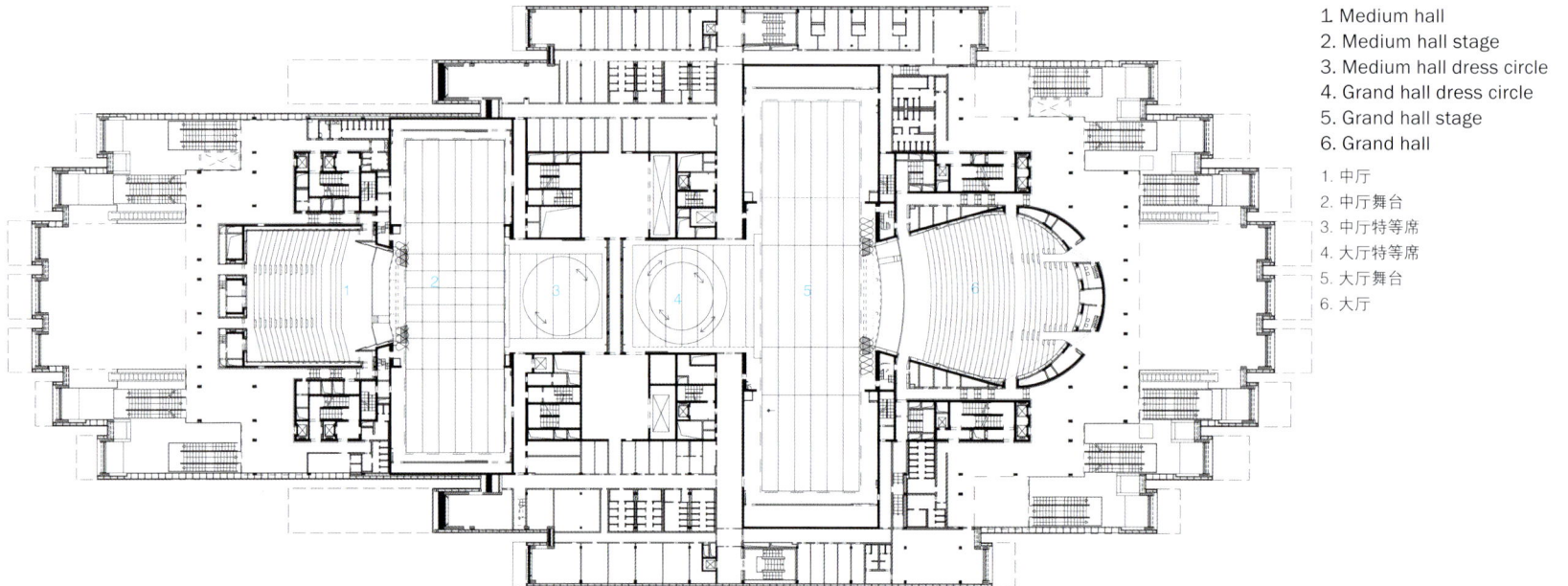

1. Medium hall
2. Medium hall stage
3. Medium hall dress circle
4. Grand hall dress circle
5. Grand hall stage
6. Grand hall

1. 中厅
2. 中厅舞台
3. 中厅特等席
4. 大厅特等席
5. 大厅舞台
6. 大厅

1

2

1. Natural light passes through the glass façade into the interior
2. Internal pathways
3. Wall structure of the theatre

1. 自然光线透过玻璃外层洒入剧院内部
2. 剧院内部通道
3. 大剧院墙体结构

3

1. Grand hall auditorium
2. View of the grand hall from the balcony
3. Medium hall auditorium
4. Rehearsal hall
5. Front hall to rehearsal

1. 大剧场观众席
2. 从二楼俯看大剧场
3. 中剧场观众席
4. 排练厅
5. 剧场前厅

玛戈特和比尔·温斯比尔歌剧院

The Margot and Bill Winspear Opera House

Location: Dallas, USA **Designer:** Foster + Partners **Completion date:** 2009 **Photos©:** Iwan Baan, Tim Hursley

项目地点：美国，达拉斯 设计师：福斯特建筑事务所 完成时间：2009年 摄影师：伊万·巴恩、蒂姆·赫斯利

The Margot and Bill Winspear Opera House is engineered specifically for performances of opera and musical theatre with its stages equipped for performances of ballet and other forms of dance.

A 21st century reinterpretation of the traditional "horseshoe" opera house, the Winspear Opera House's principal performance space, the Margaret McDermott Performance Hall, will seat 2,200 (with capacity up to 2,300) and will feature retractable screens, a spacious fly-tower and variable acoustics. The Winspear Opera House will also include the Nancy Hamon Education and Recital Hall, a space that can be used for smaller performances seating audiences up to 200, as well as classes, rehearsals, meetings and events.

The opera house's principal entrance features the 60-foot Annette and Harold Simmons Signature Glass Façade that wraps three-quarters of the way around the building, creating a transparency between the opera house and the surrounding Performance Park. The transparent façade provides dramatic views of the Margaret McDermott Performance Hall, which will be cladded in vibrant red glass panels. From within the Winspear Opera House, the Simmons Glass Façade provides a sweeping view of the skyscrapers of downtown Dallas that line the northern edge of the Performance Park.

To enhance the connection with the Park, an 84-foot wide section of the glass façade will be retractable to a height of 23 feet, literally opening up the Grand Lobby, Cafe and Box Circle-level restaurant to Sammons Park.

Radiating from the Winspear Opera House on all sides, the sky canopy provides shade over three acres of the Sammons Park, creating new outdoor spaces for visitors to gather and relax. The glass solar canopy's louvers are arranged at fixed angles following the path of the sun, calculated to provide optimal shade for the outdoor spaces throughout the day, as well as preventing direct sunlight from hitting the Simmons Glass Façade during the warmest months of summer.

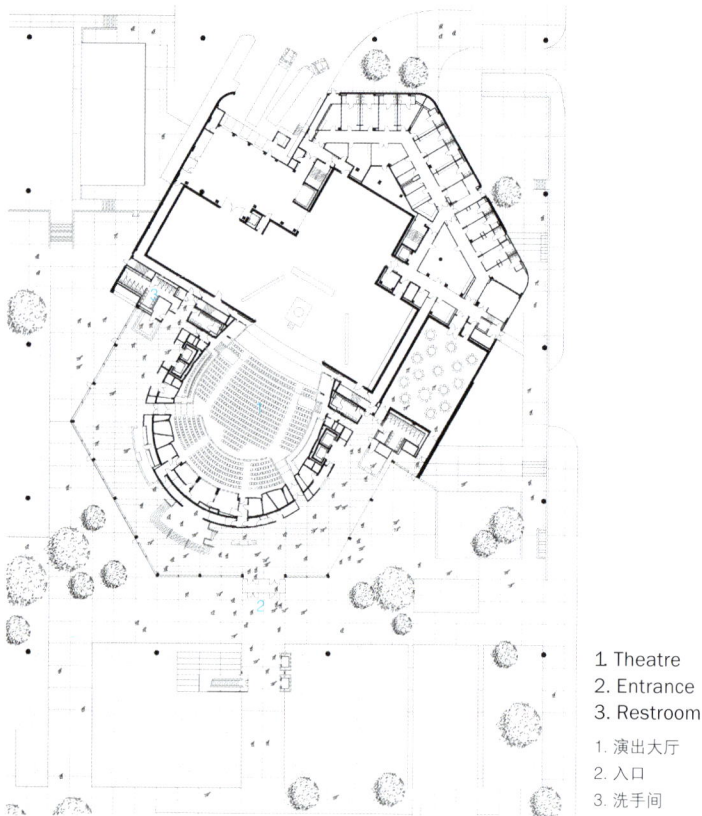

1 Theatre
2 Entrance
3 Restroom

1. 演出大厅
2. 入口
3. 洗手间

玛戈特和比尔·温斯比尔歌剧院特别为歌剧和音乐剧表演设计，其舞台还配备了芭蕾和其他舞蹈表演所需的设施。

作为一个21世纪的传统马蹄铁形剧院——温斯比尔歌剧院的主要表演空间——玛格丽特·麦克德莫特表演厅拥有2,200（最多2,300）个坐席，以可伸缩式屏幕、宽敞的飞行塔和可调节的音响效果为特色。温斯比尔歌剧院还将包含南希·哈蒙教育和演奏厅——一个可容纳200人的小型表演空间，演奏厅还能进行教学和排练，举办会议和活动。

歌剧院的主入口处，约18米高安妮特和哈罗德·西蒙斯玻璃墙包围了建筑的四分之三，在歌剧院和表演公演之间形成了透明的屏障。透明外墙提供了包围在红色玻璃板中的玛格丽特·麦克德莫特表演厅引人注目的景象。在温斯比尔歌剧院内部，透过西蒙斯玻璃墙可以遥望表演公园北侧、达拉斯市中心的高楼大厦。

为了与公园紧密相连，约25米宽的玻璃幕墙可以缩成7米高，让大厅、咖啡厅和餐厅面向公园开放。

温斯比尔歌剧院的穹顶向四方辐射，覆盖了萨门斯公园1.2公顷的面积，形成了新的户外空间，供游客集会和休息。太阳能玻璃穹顶的栅格将追踪太阳的轨迹，呈一定的角度，在白天估算出最佳的阴凉空间，也防止了夏天太阳直射到西蒙斯玻璃墙上。

1. Exterior night view
2. Exterior, the red architecture in water mirror
3. View showing the Margot and Bill Winspear Opera House
4. Front entrance of the opera house, photo©Iwan Baan

1. 建筑外部夜景
2. 红色灯光中，建筑倒映在水中
3. 玛戈特和比尔·温斯比尔歌剧院
4. 歌剧院正门入口

3

4

1. The sky canopy overhead extents off of the Opera House
2. Stair and lobby

1. 歌剧院上方的顶棚延展开来
2. 楼梯和大堂

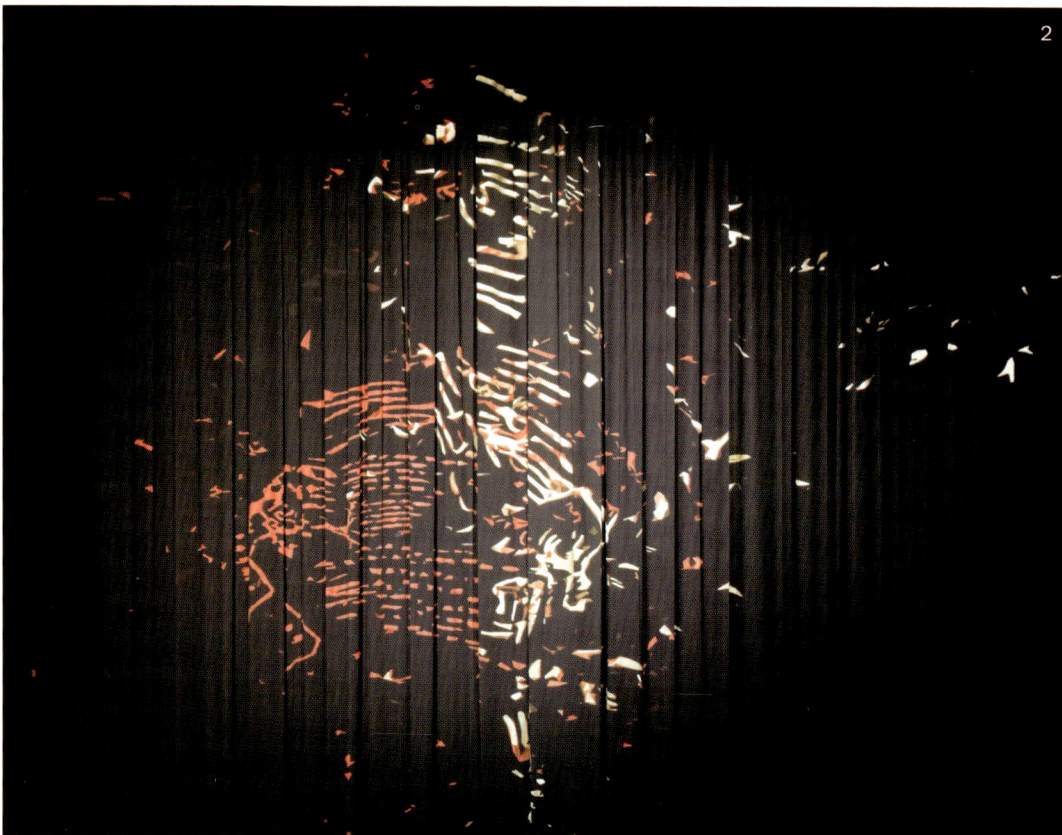

1. Chandelier of the performance hall
2. Stage curtain
3. View showing the audience seats from the stage

1. 表演厅的吊灯
2. 舞台幕布
3. 从舞台看观众席

曲线剧院 # Curve

Location: Leicester City, United Kingdom **Designer:** Rafael Viñoly Architects **Completion date:** 2008
Photos©: Rafael Viñoly Architects **Area:** 12,900 square metres **Award:** Winner of 2009 RIBA Award

项目地点：英国，莱斯特城 设计师：拉斐尔·维诺里建筑事务所 完成时间：2008年 摄影师：拉斐尔·维诺里建筑事务所 面积：12,900平方米 所获奖项：2009年英国皇家建筑师协会优胜奖

As an anchor for redevelopment of the St. George's Conservation Area in downtown Leicester, the city's new theatre, named Curve, seeks to engage community life. To fulfill this mission, Rafael Viñoly Architects turned the typical theatre configuration "inside out," exposing the production, construction, craft, and technical components of the building to the public, and integrated the performance into the life of the city itself.

The design accomplishes this goal of public engagement via a four-storey glazed curtain wall that reveals two main performance venues, a 750-seat main theatre and a 350-seat black box theatre, situated on opposite sides of the main stage and surrounded by the public ground-floor lobby. The stage, lobby, and sidewalk are all at the same level, with ample visual connections among them, thus making the theatrical performance an extension of activity on the street. Metal shutters open the stage to one theatre, to both theatres at once, or to the lobby, which allows for a wide variety of performance configurations to meet the community's diverse cultural needs.

No distinction is made between front- and back-of-house, because the stage itself can be made part of the lobby and circulation. Situated at ground level across the main lobby from the stage, double-height workshops and production spaces feature glass walls that expose production activities and make them a visible part of the performance.

Rectangular building volumes along the north and west elevations contain administrative offices, production facilities, dressing rooms, rehearsal spaces, the box office, a recording studio, a kitchen, and other support functions. Circulation balconies at upper levels overlook the foyer and allow for physical and visual connections among staff, performers, and the audience to activate a dramatic, engaging public space.

"Curve is an extraordinary contribution to the regeneration of Leicester," says Rafael Viñoly. "This project could not have been if it weren't for the vision of the people involved. They were interested in this notion of a theatre being an inside-out experience, something in which the production has as much value as the performance itself."

作为莱斯特市中心圣乔治保护区重建工程的标志性建筑，曲线剧院试图融入社区生活。为了实现这一任务，拉斐尔·维诺里建筑事务所颠覆了传统的剧院设计，将建筑的制作、建设、工艺和技术公诸于众，并且在其中融入了城市生活元素。

设计通过四层玻璃幕墙将两个主表演厅——750坐席的主厅和350坐席的实验剧厅——展现在公众面前。实验剧院位于主舞台的对面，周围环绕着公共大厅。舞台、大厅和走道都设在同一水平上，中间拥有足够的视觉连接点，让剧院表演一直延伸到街道上。金属百叶窗让舞台向剧院或者大厅开放，保证了不同的表演形式，以满足社区的文化需求。

剧院的前后没有区别，因为舞台也可以是大厅和走道的一部分。一楼的工作室和制作室从舞台穿越了大厅，以玻璃墙为特色，让创作活动成为表演的一部分。

西北向的矩形建筑空间包含行政办公室、创作设施、更衣室、排练空间、售票处、录音室、厨房和其他辅助设施。上层的圆形包厢俯瞰着大厅，让工作人员、表演者和观众从实际和视觉上连接起来，打造了一个戏剧化、迷人的公共空间。

拉斐尔·维诺里称："曲线剧院为莱斯特的复兴做出了重大贡献。如果没有考虑到人的参与，项目将大不相同。人们对剧院所提供的由内而外的体验而感兴趣。制作过程与表演同样重要。"

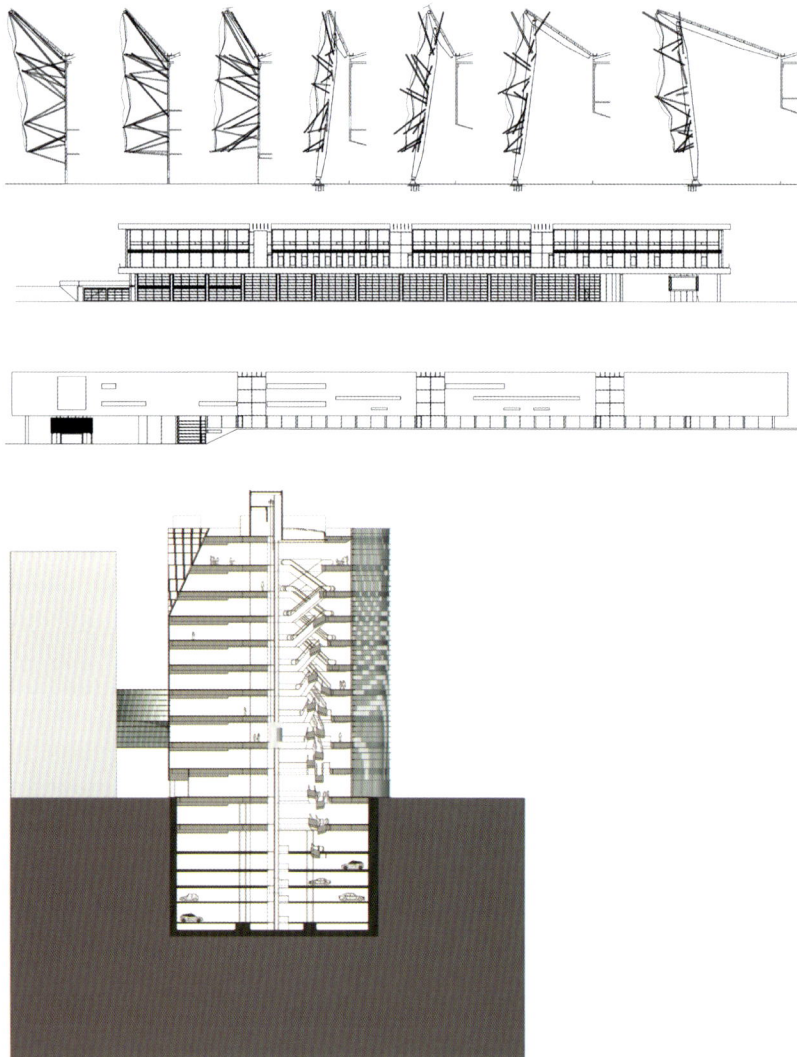

1. The theatre is surrounded by trees and buildings
2. At night, red light comes through the windows
3. The façade of the theatre

1. 被树木和建筑群包围的剧院
2. 夜晚剧院内部透出红色灯光
3. 剧院外立面

1

2

1, 2. The theatre's façade looks like a metal shutter

1、2. 剧院外立面像一个巨大的金属百叶窗

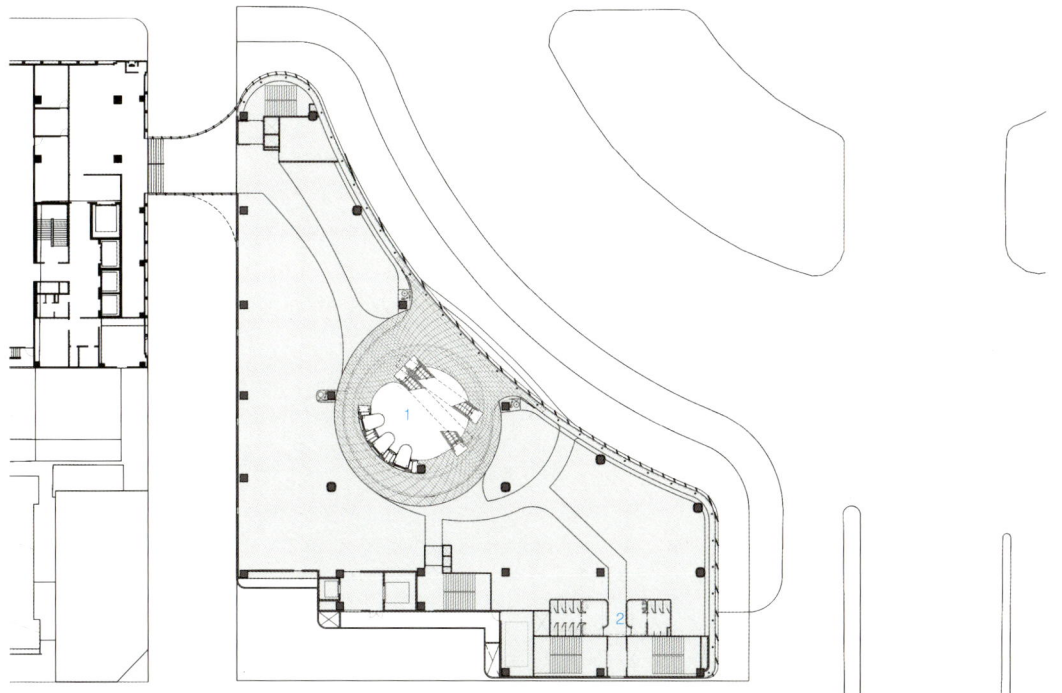

1. Black box theatre
2. Toilets

1. 实验剧场
2. 洗手间

1. Interior red walls of the theatre
2. Sunlight passes through the glass wall into the interior red walls
3. Ticket office on the ground floor
4. View showing the outside street through the window of the ground floor

1. 剧院内部红色墙体
2. 阳光透过玻璃外墙投射到室内红色墙体上
3. 一层售票处
4. 透过一层大厅的窗体可以看到外部街道

1. Stage
2. Auditorium

1. 剧院舞台
2. 观众席

新皇家剧场 # The New Royal Playhouse

Location: Copenhagen, Danmark **Designer:** Lundgaard & Tranberg Arkitekter **Completion date:** 2007
Photos©: Jens Lindhe **Awards:** The Association For Preserving The Beauty Of The Capital City,
Diploma 2007/The Danish Lighting Award 2008/RIBA European Award 2008/The Nordic Lighting
Award 2008/Copenhagen Municipality 2008/Sustainable Concrete Prize 2009

项目地点：丹麦，哥本哈根 设计师：朗德加尔德&特朗伯格建筑事务所 完成时间：2007年 摄影师：延斯·林德 所获
奖项：2007年首都美丽维护协会证书；2008年丹麦灯光奖；2008年英国皇家建筑师协会欧洲奖；2008年哥本哈根
市政奖；2009年可持续混凝土建筑奖

1. Stage
2. Auditorium
3. Toliets

1. 舞台
2. 观众席
3. 卫生间

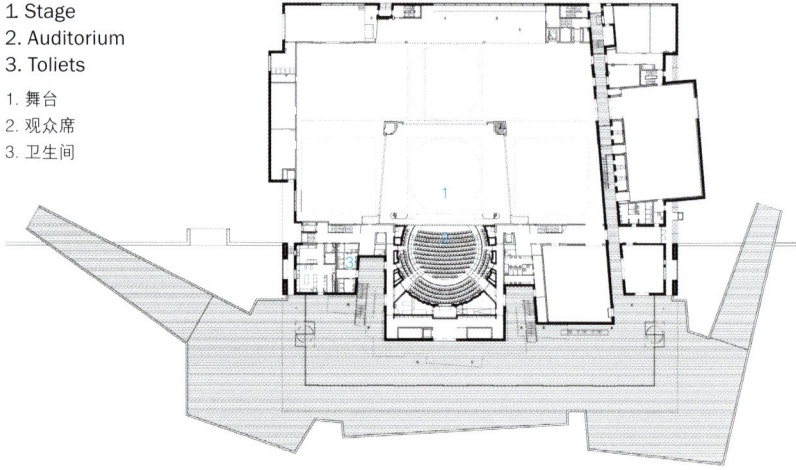

1. Situated on the seaside, the playhouse acts as an anchor for this meeting of city and sea
2. Long and elegant exterior of the playhouse
3. A night view of the playhouse

1. 位于海滨的剧场是城市与海洋的交叉点
2. 细长而优雅的剧场外观
3. 剧场夜景

The Royal Playhouse is located at south of where Sankt Annæ Plads, one of Copenhagen's finest urban spaces, and meets the harbour. Here the elegant, elongated urban room meets the waterfront, giving way to view that follow the harbour north to the distant sound. The playhouse acts as an anchor for this meeting of city and sea, revealing and reinforcing the existing urban spatial qualities.

The primary theatre spaces are housed in the massive masonry scene building, which echoes the surrounding historic harbour warehouses. The dark brick of the exterior is drawn deep into the interior, creating a grotto-like universe of textures and dramatic lighting. The magic of the theatre awakens upon encountering the raw masonry walls, which enclose the various scenes and provide the backdrop for the foyer. Here there is no open transparency, here the eye meets niches, balconies, and stairs – and only seldom, isolated views reveal the secrets of the quiet, powerful, masonry form. A combination of strafing and muted artificial light reveals the distinctive and cragged character of the individual bricks – the fountainhead of the theatre's intense, raw energy. The expansive, projecting upper level, containing artists functions, workspaces, and administration, is borne by full-storey steel trusses and is sheathed in varying hues of green glass. Daylight and views of the harbour flood in during the day, and at twilight a transformation occurs – the upper level pulses with colour and light, presenting a nightly show on the urban stage, signalling life and creative activity.

The primary organisational concept of the playhouse imparts compactness and minimises the spread of the building, thereby reducing the distances between the many functions and providing an easily understood spatial layout. The complex interrelation of spaces and uses was generative in the development of the buildings concept and disposition. The design of the playhouse and way it functions are inseparable.

At the heart of the playhouse is a circular, grotto-like auditorium, seemingly carved out of the masonry mass of the scene building. The main stage is constructed of striking, staggered masonry walls, providing the necessary acoustic environment with a reverberation time of one second. The specially designed red velour chairs follow the room's concentric geometry, creating an intimate relationship between actors and spectators, where every sight, sound, and breath is shared.

皇家剧场位于哥本哈根最好的城市空间之———圣安娜广场与港口的交叉处之南。细长而优雅的城市空间与海滨十分相称，可以遥望港口到海峡的景色。剧场是城市与海洋的交叉点，展示和强化了现有的城市空间和品质。

主要的剧院空间处在巨大的石造建筑之中，与港口古老的仓库十分相称。外墙的暗色砖石深入到室内，形成了洞穴感的纹理和灯光效果。剧院的魔力唤醒了粗糙的石墙。石墙围住了各种情景，也为大厅提供了背景。这里没有开放和透明，满眼是壁龛、包厢和楼梯，石造结构安静、有力又充满了神秘气息。扫射和柔和的灯光结合在一起，展现了石块与众不同而凹凸崎岖的特征，也体现了剧院紧张感和能量的根源。

广阔而突出的上层空间包含艺术功能、工作室和行政空间，由钢桁架支撑，覆盖在不同色调的绿色玻璃之下。白天，日光和海港的景色渗入剧院之中；傍晚，变化开始了，上层空间的色彩和灯光在跳动，在城市舞台上演了一场夜间剧，引导着生活和创意活动的进行。

剧场的主要组织概念强调了紧凑，最小化了建筑的幅度，从而减少了各个功能区之间的距离，形成了更简单的空间布局。空间和功能的相互关系在建筑理念和配置开发中具有衍生价值。剧场的设计与其功能密不可分。

剧场的中心是一个圆形的洞穴感礼堂，仿佛从建筑中挖出的一样。主舞台由错列的石造墙壁建造而成，提供了必要的音响环境，混响时间达1秒钟。特别设计的红色天鹅绒座椅遵循了空间的几何轴线，增进了观众和演员之间的联系，让声、色、呼吸都得到了共享。

1

2

1. People could take a rest and have a view of the sea in front of the playhouse
2. Sea terrace connecting the playhouse and the city
3. Sea terrace outside the playhouse
4. Stairs connecting each levels

1. 人们在剧场前休憩观海
2. 连接剧场与城市的海上平台
3. 剧场外滨水平台
4. 剧场内连接各层的楼梯

1

2

1. Through the windows of the restaurant, you will have a panoramic view of the city at night
2. The sculpture in the corridor is quite artistic
3. Different levels have different views

1. 透过餐厅玻璃窗将夜晚城市美景尽收眼底
2. 走廊上的雕塑平添艺术气息
3. 内部各层风景各不同

3

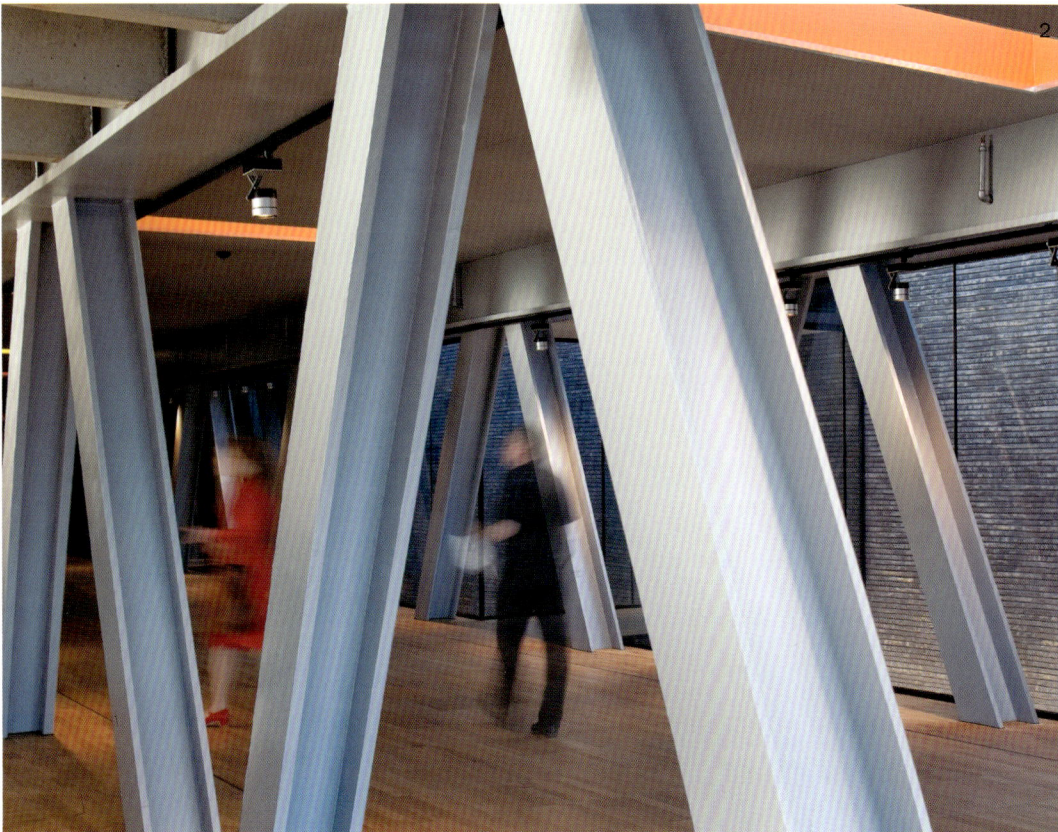

1. Seascape studio
2. A-shaped internal structural support
3. Seats in the performance hall

1. 海景工作室
2. 内部A形支撑结构
3. 演出大厅观众坐席

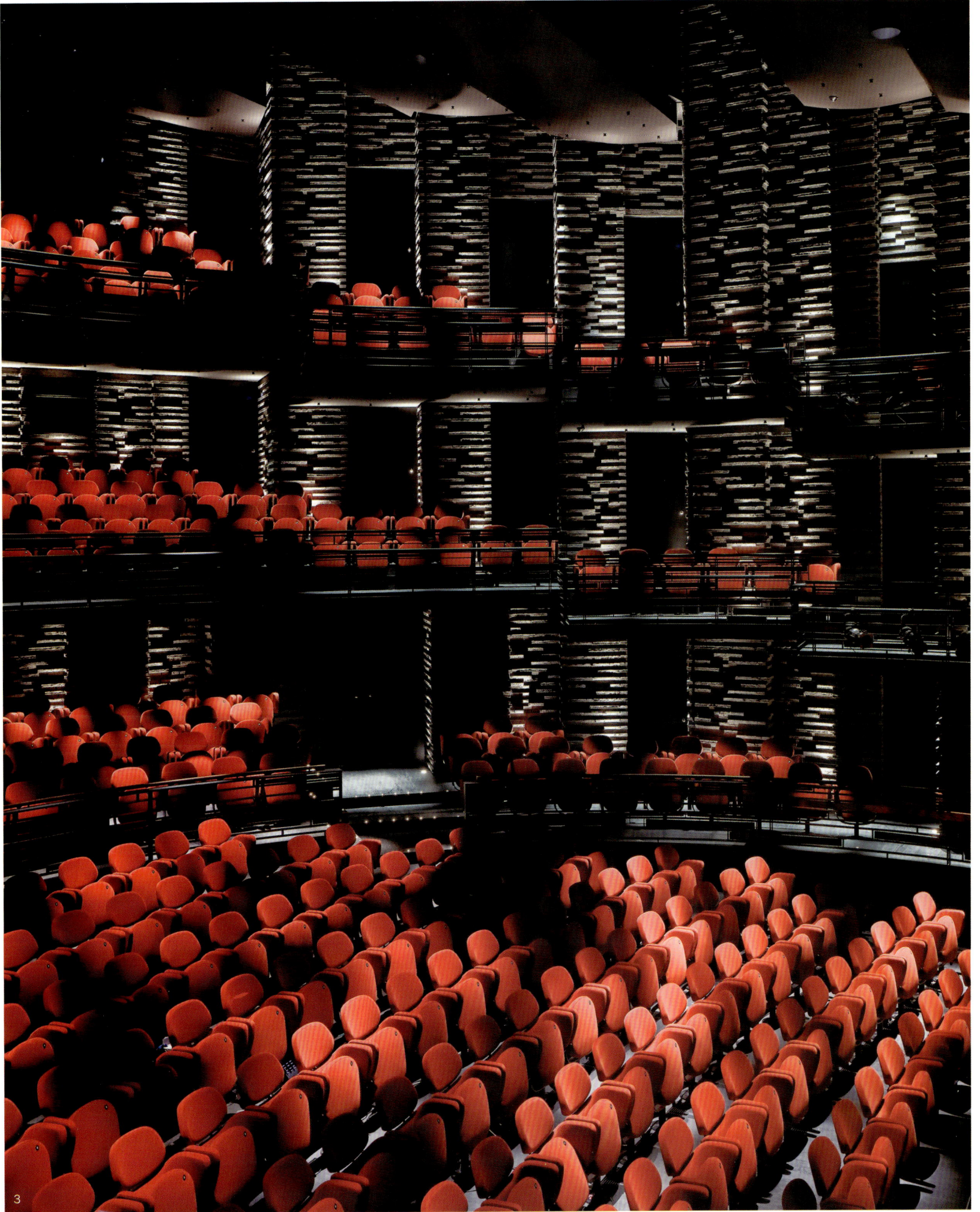

3

塔斯特鲁普剧院 Taastrup Theatre

Location: Taastrup, Denmark **Designer:** COBE **Completion date:** 2009 **Photos©:** Stamers Kontor **Area:** 1,150 square metres of renovation, 200 square metres of new building **Award:** Shortlisted for WAF 2010 in the Category New and Old.

项目地点：丹麦，塔斯特鲁普 设计师：COBE 完成时间：2009年 摄影师：斯塔莫尔斯·康特尔 面积：翻新1,150平方米，新建200平方米 所获奖项：入选WAF2010年度改造项目奖

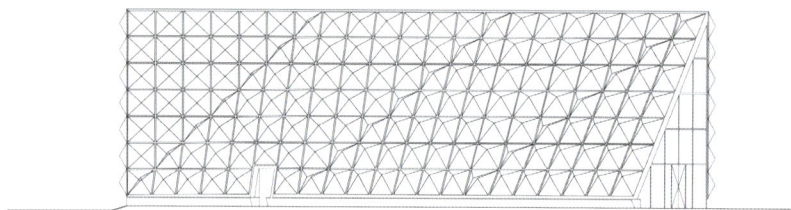

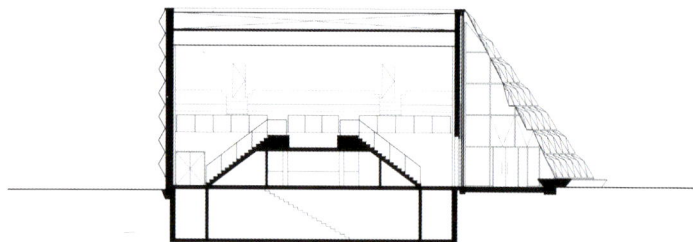

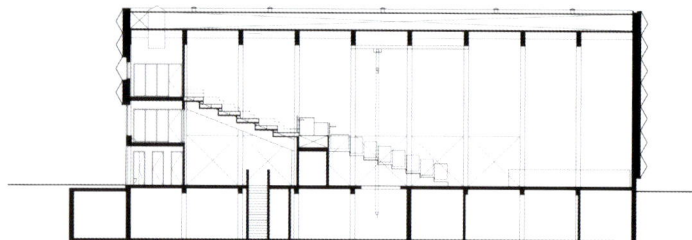

The project for the extension and renovation of Taastrup Theatre seeks to improve the communication of the building with its environment – a social housing neighbourhood. Formally COBE was commissioned to improve the energy consumption of the 1970s' local community theatre in the Copenhagen neighbourhood of Taastrup. Yet, the design team used this opportunity to improve the general appeal and functionality of the building by introducing a second (isolating) theatre curtain around the rough concrete structure. By adding this new layer in front of the existing rough concrete structure, the building was extended and opened as wide as possible towards Kjeld Abels Plads north.

The new translucent façade subtly reminds the designers of a theatre curtain about to open when the play starts. In fact, when the tickets are outsold or the show is on, red lights underneath the façade broadcast the special atmosphere of this magic moment towards the square. This new composition underlines the unity of the old building and its extension as one piece of architecture.

The façade is conceived as a translucent curtain of acrylic prisms elegantly embracing the existing building, and creating a new open foyer, arrival area, and café. A whole new spatial dimension is added to the building newly connecting the formerly enclosed interior with the outside. Now the theatre visually and spatially connects to both Kjeld Abels Plads north and the protected green courtyard south of the theatre. With this overall concept, the theatre now has the possibility to involve the surroundings by exposing its activities. The façade is assembled from acrylic prisms. The material is very robust and because of its shape it has an exciting effect at day and night time, when the light from inside will glow out to the surroundings. The prisms vary from totally clear and transparent over translucent to opaque. Because of many elements of irregular shapes and the variance in translucency, the façade is a beautiful play of shadows and reflections.

The façade of the theatre includes a number of gates creating connection and access between the foyer and the theatre space. These gates can be opened or closed depending on the arrangement. So the theatre has the possibility to work as one coherent floating space, or separately as a theatre space and a foyer.

塔斯特鲁普剧院扩建和翻新项目旨在提升建筑与周边环境——一个住宅社区的交流。起先，COBE只是受委托为这个建于19世纪70年代的社区剧院改善能源消耗状况。然而，设计团队利用这一机会改善了建筑的整体设施和功能性，在粗糙的混凝土结构外引入了一个独立的剧院幕墙。通过添加这层新外壳，建筑得以扩展，面向杰尔德·阿贝斯广场北侧的立面得到了最大化。

新的透明外墙让设计师想起了剧目即将开始拉开的幕布。事实上，当剧票售完或表演开始时，外墙下方的红灯将向广场放送这个奇妙时刻的独特氛围。这个新结构强化了旧建筑与扩建工程的统一性。

由亚克力棱镜组成的透明外墙优雅地包围着原有的建筑，形成了一个新的开放式大厅、到达区和咖啡厅。建筑增添了一个全新的空间维度，使封闭的室内和外部紧密相连。剧院从视觉和空间上与杰尔德·阿贝尔斯广场北侧和南侧的绿色庭院连接起来。这个全局概念让剧院通过展示自己的活动与周边结合在一起。亚克力材料十分牢固，由于造型的缘故，玻璃外墙在日夜拥有不同的光影效果，内部的灯光将会散发到周边的环境里。棱镜从透明过渡到不透明，直到不透明。由于不规则造型和透明度的变化，外墙的光影效果十分迷人。

剧院的外墙包含许多大门，与大厅和剧院空间相连。这些门可以根据布局开放和闭合，使剧院既可以形成一个连贯的空间，又能使剧院空间和大厅分隔开来。

1. Night view   1. 夜景
2. Dusk view   2. 黄昏中的建筑

3

4

1. Entrance
2. Resting
3. Reception
4. Auditorium
5. Bathroom

1. 入口
2. 休息区
3. 前台
4. 观众席
5. 洗手间

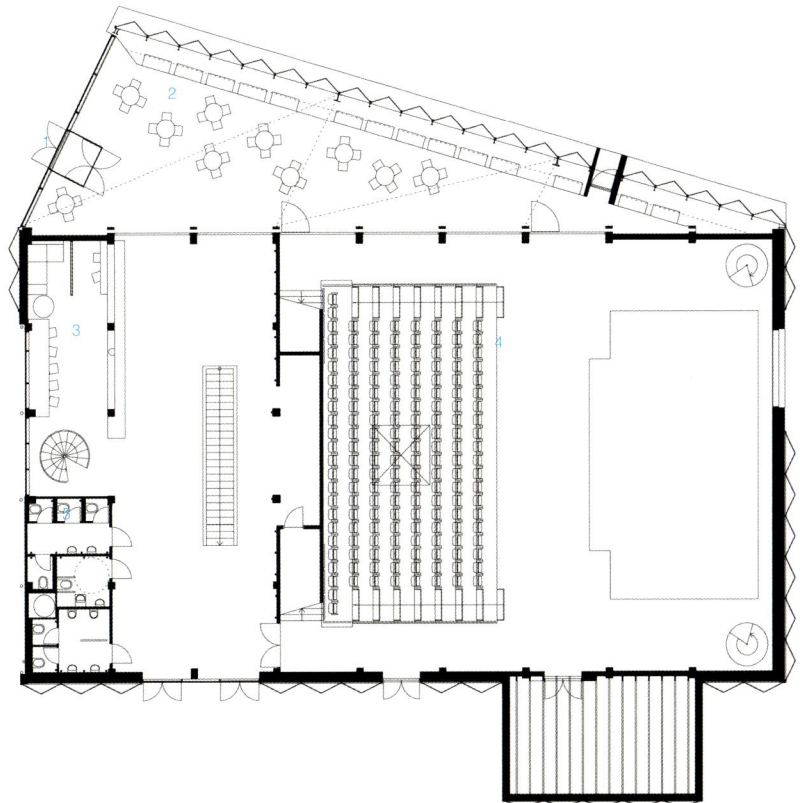

1. General view
2. Façade
3, 4. Facade detail

1. 建筑全景
2. 建筑外观
3、4. 外墙细部

2

1, 2. Lounge and café

1、2. 休息室和咖啡厅

1

1. Lounge and café
2. Reception

1. 休息室和咖啡厅
2. 前台

2

阿格拉剧院 Agora Theatre

Location: Lelystad, The Netherlands **Designer:** UN Studio and Van Berkel & Bos Architectuurbureay
Completion date: 2007 **Photos©:** Courtesy of UN Studio **Building area:** 2,925 square metres

项目地点：荷兰，莱利斯塔德 设计师：UN工作室，凡·伯克尔&波斯建筑 完成时间：2007年 摄影师：UN工作室提供
建筑面积：2,925平方米

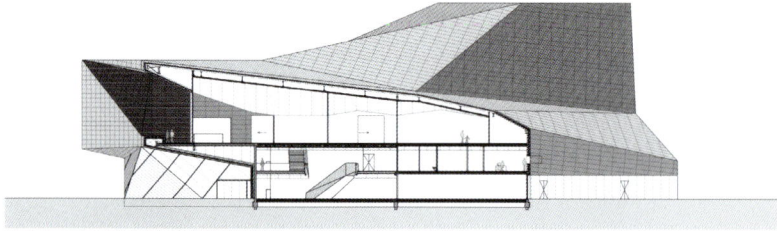

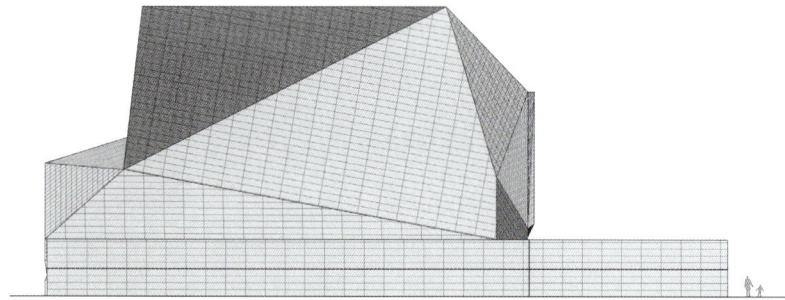

The Agora Theatre is an extremely colourful, determinedly upbeat place. The building is a part of the master plan for Lelystad by Adriaan Geuze, which aims to revitalise the pragmatic, sober town centre. The theatre responds to the ongoing mission of reviving and recovering the post-war Dutch new towns by focusing on the archetypal function of a theatre: that of creating a world of artifice and enchantment. Both inside and outside walls are faceted to reconstruct the kaleidoscopic experience of the world of the stage, where you can never be sure of what is real and what is not. In the Agora Theatre drama and performance are not restricted to the stage and to the evening, but are extended to the urban experience and to daytime.

The typology of the theatre is fascinating in itself, but the architect, Ben van Berkel, who has a special interest in how buildings communicate with people, aims to exploit the performance element of the theatre and of architecture in general far beyond its conventional functioning. As he recently stated: "the product of architecture can at least partly be understood as an endless live performance. As the architectural project transforms, becomes abstracted, concentrated and expanded, becomes diverse and evermore scaleless, all of this happens in interaction with a massive, live audience. "

The facetted outlines of the theatre have a long history in the work of UN Studio and Van Berkel & Bos Architectuurbureau before that. In this case, the envelope is generated in part by the necessity to place the two auditoriums as far apart from each other as possible for acoustic reasons. Thus, a larger and a smaller theatrical space, a stage tower, several interlinked and separate foyers, numerous dressing rooms, multifunctional rooms, a café and a restaurant are all brought together within one volume that protrudes dramatically in various directions. This facetted envelope also results in a more even silhouette; the raised technical block containing the stage machinery, which could otherwise have been a visual obstacle in the town, is now smoothly incorporated. All of the façades have sharp angles and jutting planes, which are covered by steel plates and glass, often layered, in shades of yellow and orange. These protrusions afford places where the spectacle of display is continued off-stage and the roles of performers and viewers may be reversed. The artists's foyer, for instance, is above the entrance, enabling the artists to watch the audience approaching the theatre from a large, inclined window.

阿格拉剧院色彩鲜艳，充满了活力，是莱利斯塔德总体规划的一部分。城市的总体规划旨在复兴务实而严肃的市中心。剧院的任务是通过其功能复兴战后的荷兰新城，从而打造一个奇幻而令人着迷的世界。墙壁的内外两侧都进行了切面处理，为舞台营造出万花筒般的感觉，令人真假难辨。阿格拉剧院所上演的剧目和表演不局限于舞台和夜晚，可以一直延伸到都市生活和白天里。

剧院的魅力本应在于其上演的剧目，但是本·凡·伯克尔十分注重建筑与人们的交流，试图开发剧院的演出元素，使其与建筑结合在一起，形成更大的能量。正如他所说："建筑的制作可以被看作是永无止境的生活演出。建筑的改造、抽象、集中和扩张都在一大群观众面前进行。"

剧院的切面造型拥有悠久的历史。剧院的外壳根据空间需要而形成，出于音响的原因，两个礼堂隔得越远越好。此外，剧院还要包括一大一小两个剧场、一个舞台塔、一些相连和独立的大厅、许多更衣室、多功能室、一个咖啡厅和一个餐厅，它们分别处在不同的方位。切面外壳让剧院的轮廓更加平缓，垫高的技术楼里放置着舞台机械装置，与城市结合在一起。外墙全部拥有尖锐的棱角和凸起的平面，上面覆盖着层叠的钢板和玻璃板，呈黄色和橙色。这些凸起让表演在舞台之下得以延续，颠倒了表演者和观众的角色。例如，艺术家的休息室位于入口的上方，他们可以看到观众从一个巨大而倾斜的窗户走进剧院。

1. A general view of the theatre   1. 剧院外观全景
2. The façade is yellow and orange   2. 呈黄色和橙色的剧院外壳

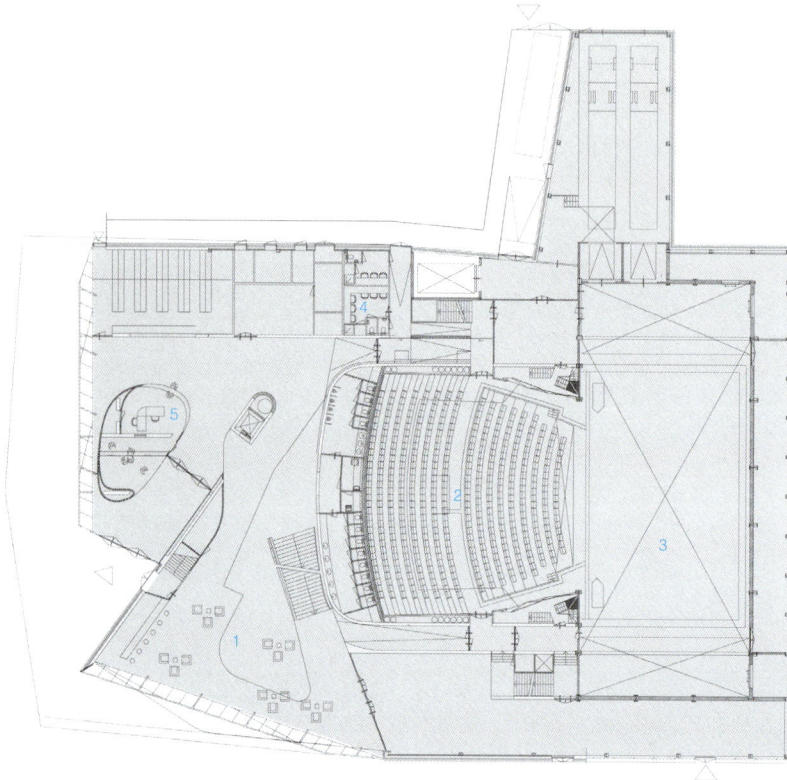

1. Resting area
2. Seating
3. Stage
4. Dressing room
5. Reception of the ground floor

1. 休息区
2. 坐席
3. 舞台
4. 更衣室
5. 一楼前台

1. Resting area
2. Staircase
3. Ceiling
4. Pathway

1. 休息区
2. 楼梯
3. 天花板
4. 通道

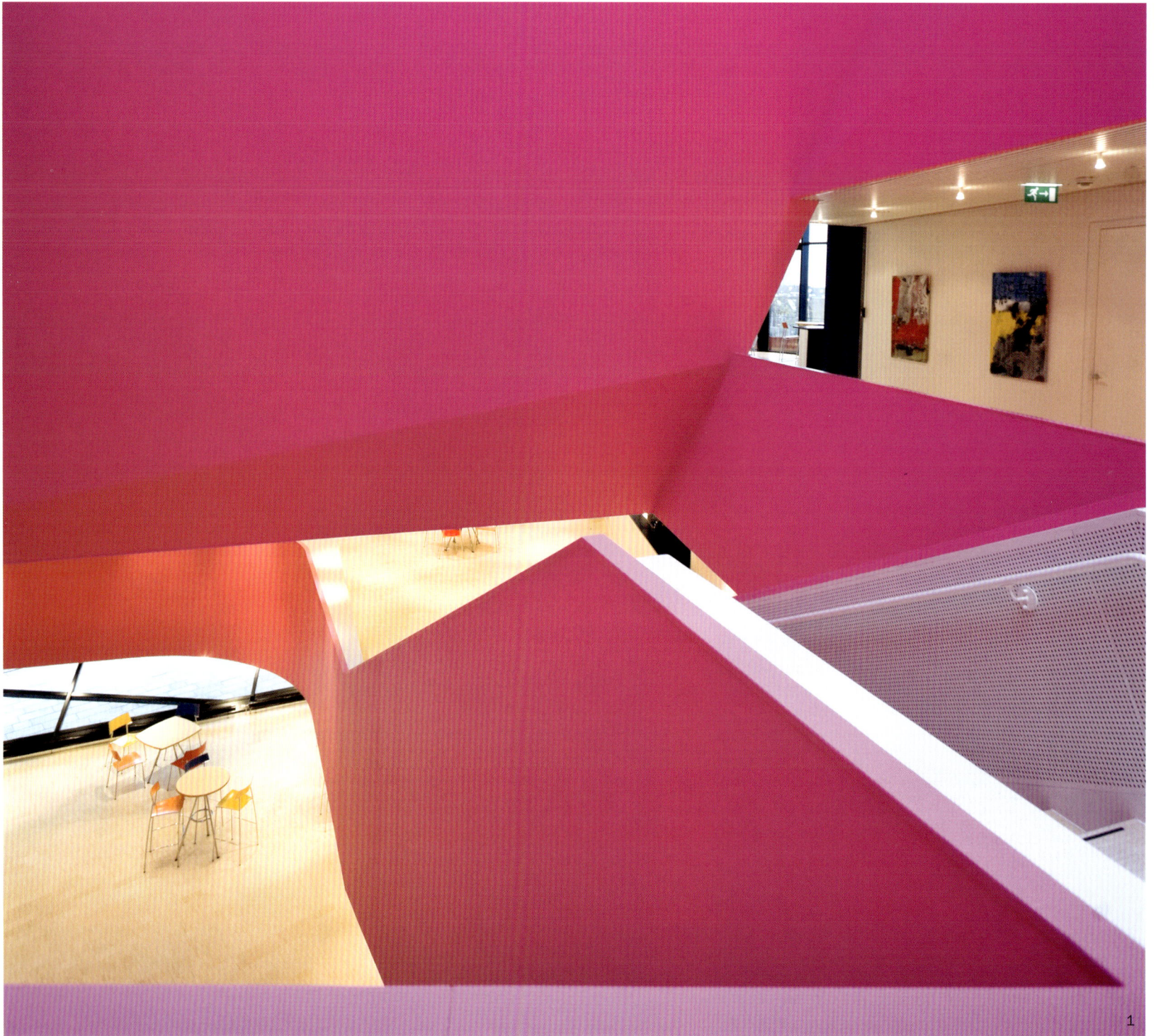

1. The facetted outlines of the theatre echo with the façade
2. Details of the ceiling

1. 与建筑外壳相呼应的切面空间造型
2. 天花板细部

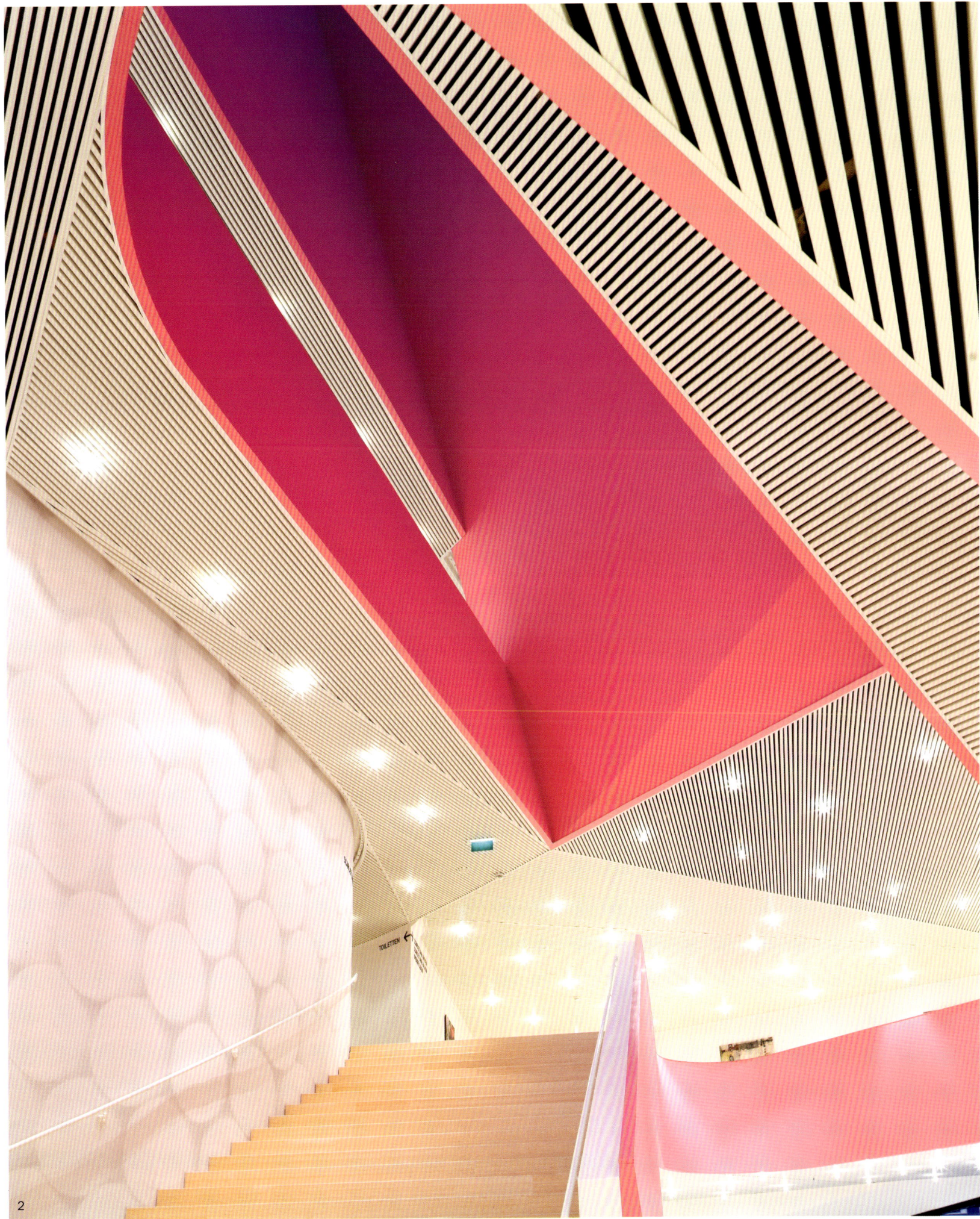

2

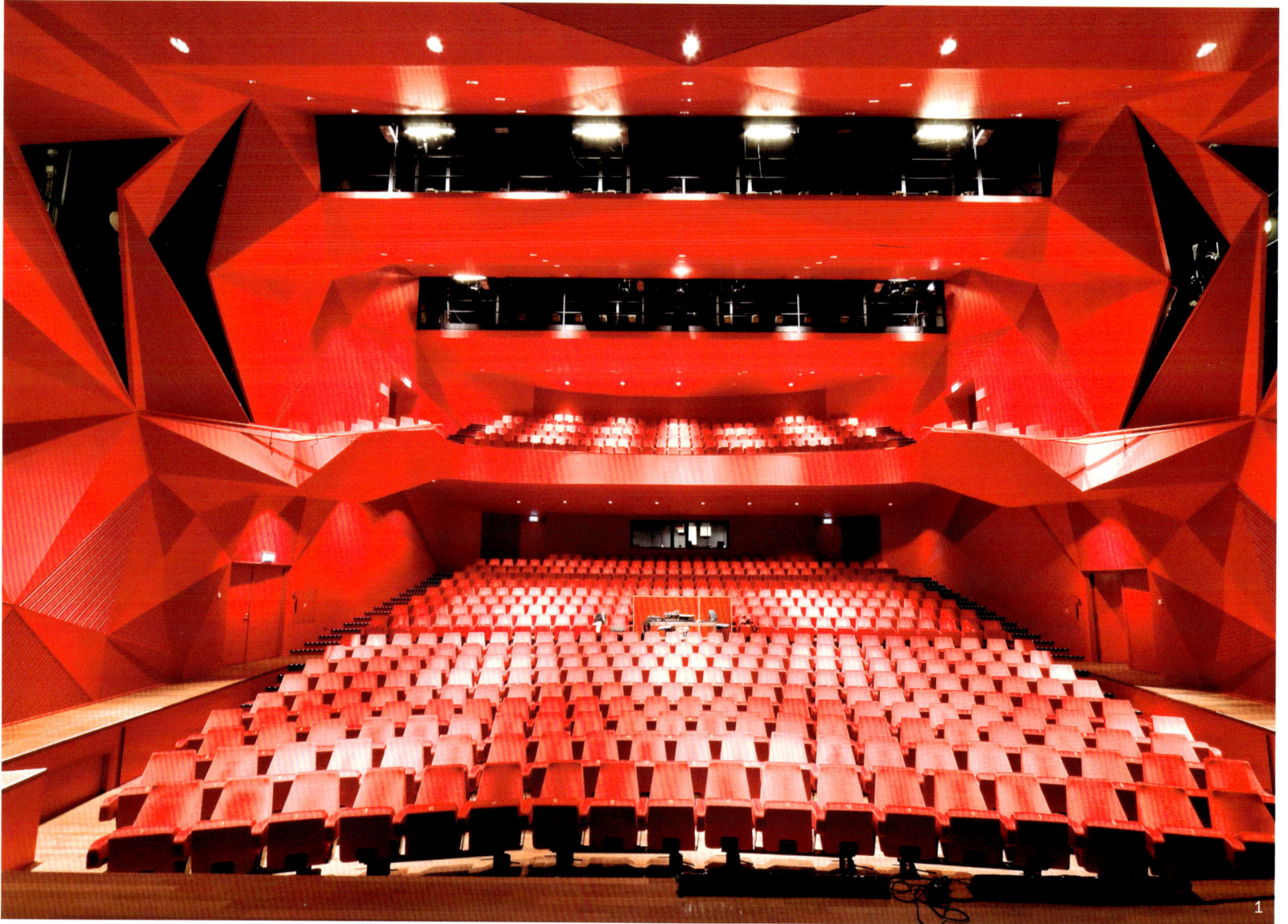

1

2

1. Auditorium
2. Auditorium side view
3. Upper level audience seats (VIP seats)

1. 剧院观演大厅观众坐席
2. 观众席侧景
3. 二层观众坐席（VIP坐席）

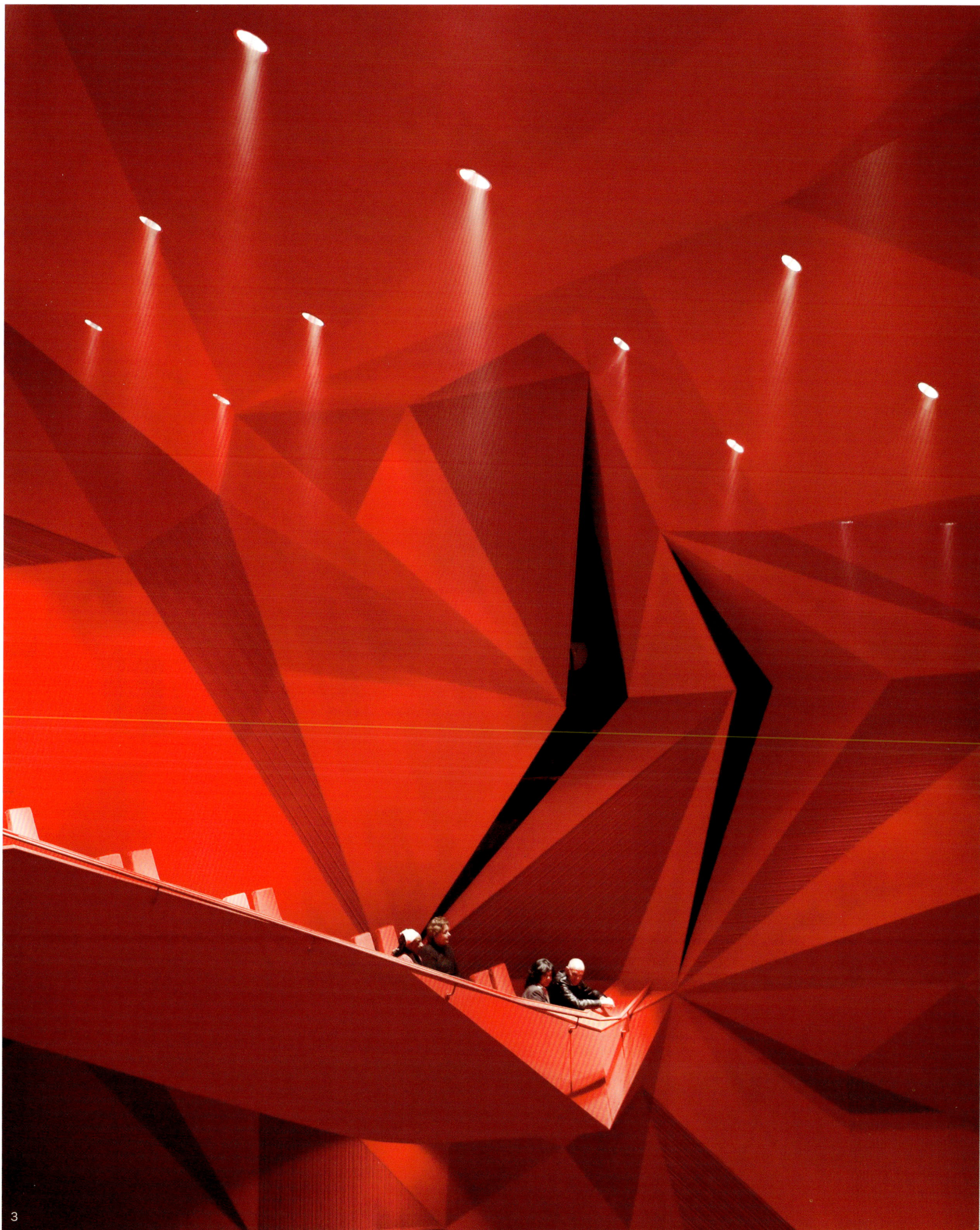

3

丹尼斯·佩雷蒂尔剧院　# Theatre Denise Pelletier

Location: Quebec, Canada **Designer:** Saia Barbarese Topouzanov Architectes **Completion date:** 2009
Photos©: Frederic Saia, Vladimir Topouzanov, Michel Dubreuil **Construction area:** 4,560 square
metres **Award:** Awards of Excellence in Architecture, Ordre des Architectes du Québec, 2011

项目地点：加拿大，魁北克 设计师：萨亚·巴巴里斯·托普扎诺维建筑事务所 完成时间：2009年 摄影师：弗雷德里
克·萨亚、弗拉基米尔·托普扎诺维、米歇尔·杜博瑞尔 建筑面积：4,560平方米 所获奖项：2011魁北克建筑师协
会优秀建筑奖

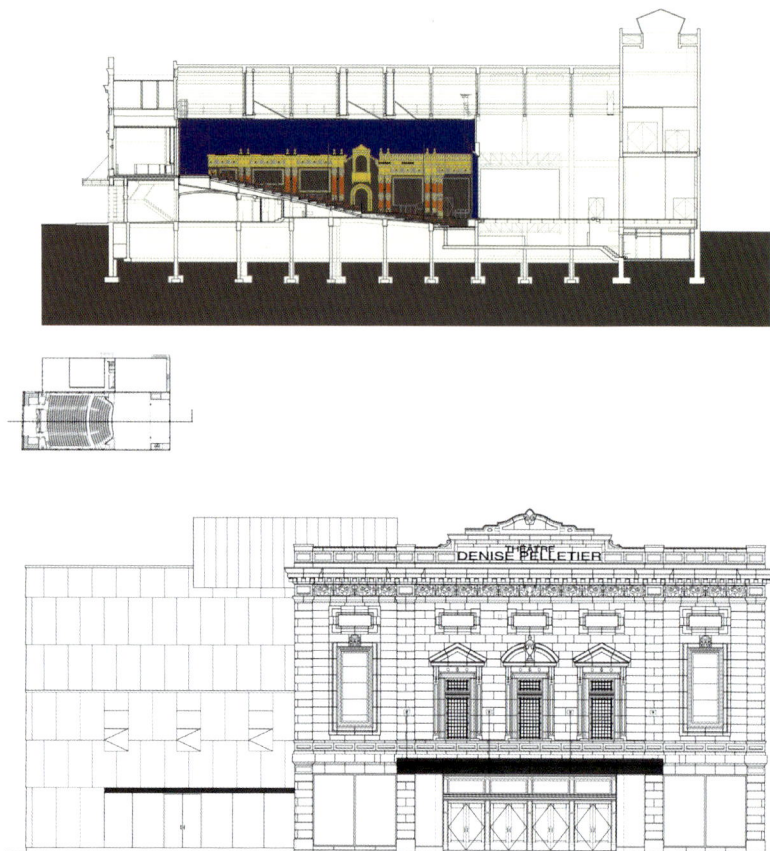

The Granada Theatre, built in 1930, is situated in the former municipality of Maisonneuve, at the northwest corner of Boulevard Morgan and Rue Sainte-Catherine. Over time, the Granada changed vocation and name. The Nouvelle compagnie théâtrale acquired the building and the adjacent lot in 1976. The wear on the Granada and Théâtre Denise Pelletier caused by the passage of time from 1930 to the modifications of 1976, and from then to 2008, was obvious. Certain operational deficiencies – viewing angles and acoustics, lack of accommodation of new scenographic standards and techniques – required renovation and restoration. The mandate given to the architectural firm Saia Barbarese Topouzanov thus had two parts: first, preserve the building's patrimonial and symbolic heritage; second, refurbish the facilities, whose rundown state had become evident especially in the previous decade.

To begin with, a careful patrimonial study brought to light and prioritised elements worthy of interest, their conditions, and what needed to be done to restore them to their previous glory. The plan adopted highlighted the old features and paired them with interventions that were modern and played a role in maintaining the site's vitality.

On the exterior, the artificial stone masonry of the main façade was cleaned and restored, while fibre-cement mega-siding now covers the frontage of the 1976 expansion. The original panes from the windows on the upper floor were entrusted to a master glazier; the oak entrance doors, to a master cabinetmaker. A lighted marquee returned to the dimensions and lightness of the original, but did not copy its decoration. Where the blank west wall had awaited a new structure, a full-height window matched the old window opening on Morgan Boulevard. Its heavy oblique window head evoked the sloping floor that once, following the plans of architect Doucet, ended against the front façade. This window, bordering the west staircase, gives onto the parvis of Salle Fred Barry, on which are inscribed the black-and-white words of the installation by artist R. M. E. Goulet. Inside, the two new angled staircases, with landings and perpendicular flights, have generous dimensions. They lead audiences from the ground floor to the upper floor, from the house to the new lobby. They also bridge a gap of some eighty years, from a lower hall to an upper one, from an elaborate decor to one that plays on simple forms and colours. The common point of them is the rectangular plan and the dominant sienna colour tones.

格拉纳达剧院建于1930年，位于前梅桑纳自治区，摩根大道和圣凯瑟琳街的西北角。多年以后，格拉纳达剧院的功能和名字都发生了改变。1976年，诺威尔剧院公司买下了剧院大楼和周边的土地。2008年，格拉纳达剧院被改造成了全新的丹尼斯·佩雷蒂尔剧院。由于观看角度、音响、布景和技术的原因，剧院进行了翻新和改造。设计师的任务包含两个方面：一是保持建筑的传统；一是刷新设备，使其性能得到大幅度提高。在设计开始前，设计师对建筑进行了全面的研究，了解了具有重要价值的元素及其使用情况，明确了需要作出哪些必要的举措来对它们进行保护。设计强调了建筑的古老特征，并且于其中添加了现代元素，保持了场地的活力。

建筑外部，人造石材外墙得到了清洁和修复，纤维水泥侧线覆盖着临街的正面。专业玻璃工人负责处理上层的窗格，而橡木大门则由专业木工负责。新天幕的规模和亮度与之前的相同，但是拥有不同的装饰。西墙的空白处，落地窗与朝向摩根大道的窗户相得益彰。厚重而倾斜的窗户与倾斜的地面相搭配，遵循了建筑师杜塞的设计，一直延伸到建筑正面。这扇窗户拓宽了西面的楼梯，可以透过窗户看到天井。天井的地面上写着黑白字母，还摆放着艺术家R.-M.E.古莱特设计的装置。建筑内部，两个全新的角度楼梯有着宏大的规模。它们引领观众从一楼到二楼，从剧场到新大厅。它们还起到了年代衔接的桥梁作用，从低处的大厅到高处，采用了从西欧精致装饰到简约的形式和色彩。它们的共同点是长方形布局和赭色色调。

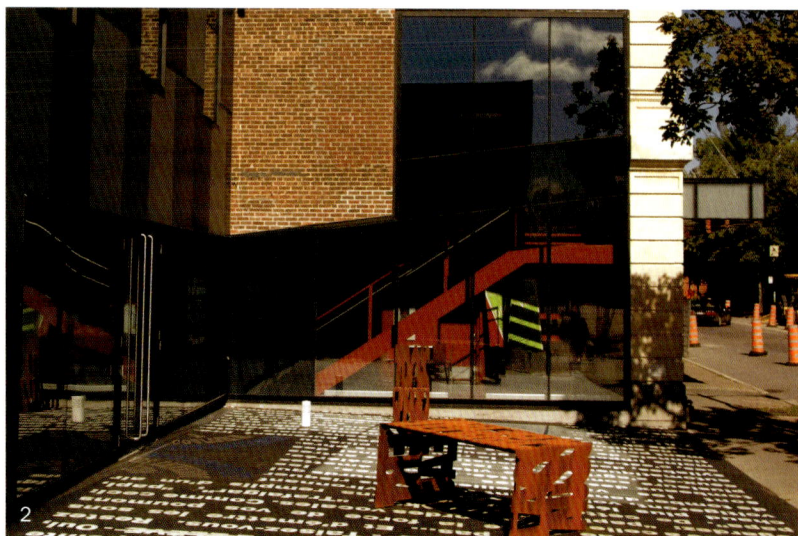

1. View of the entrance　1. 入口
2. Side view　2. 建筑侧面
3. Main facade view　3. 建筑外观
4. View of the entrance　4. 入口

3

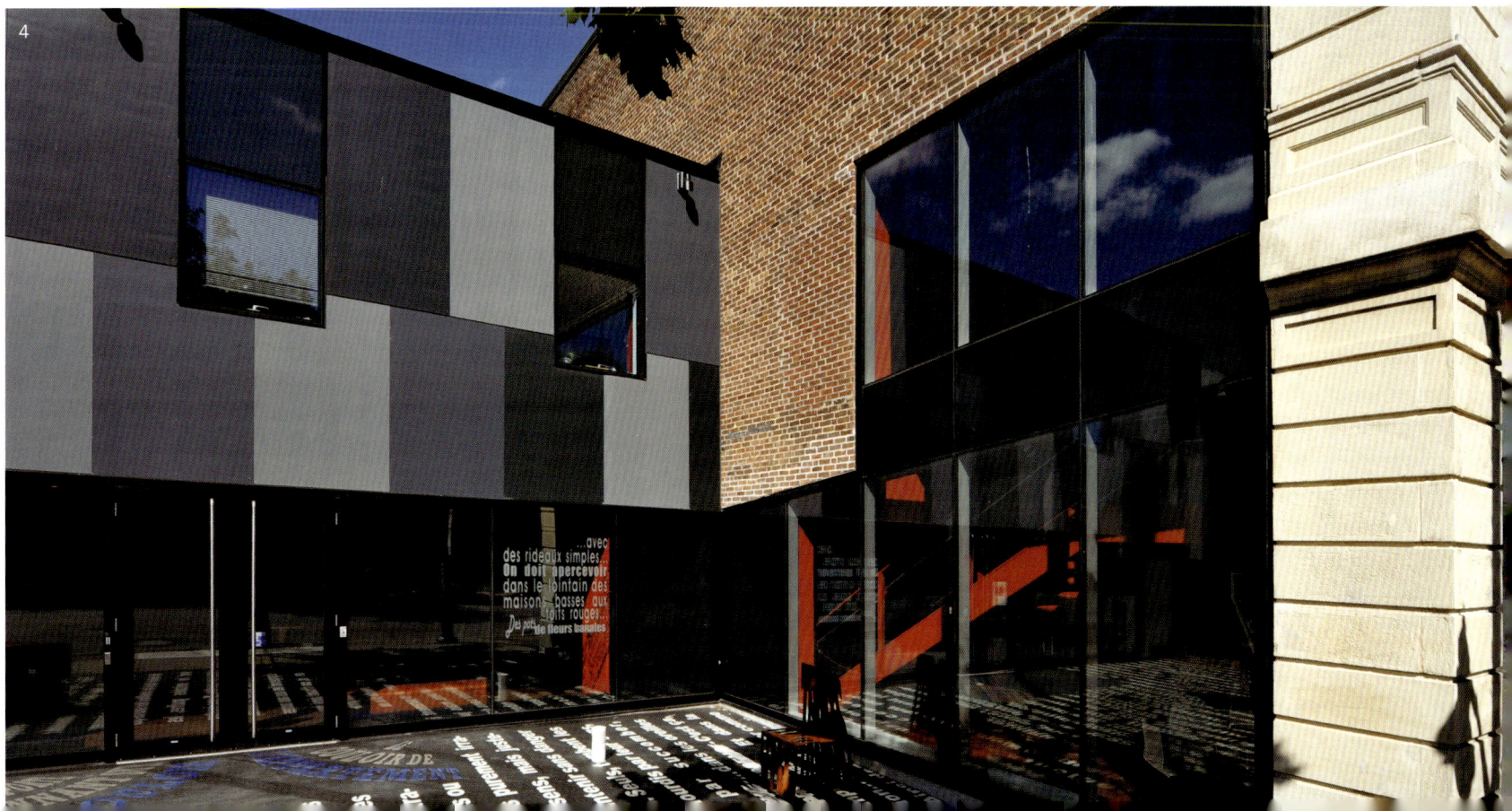

4

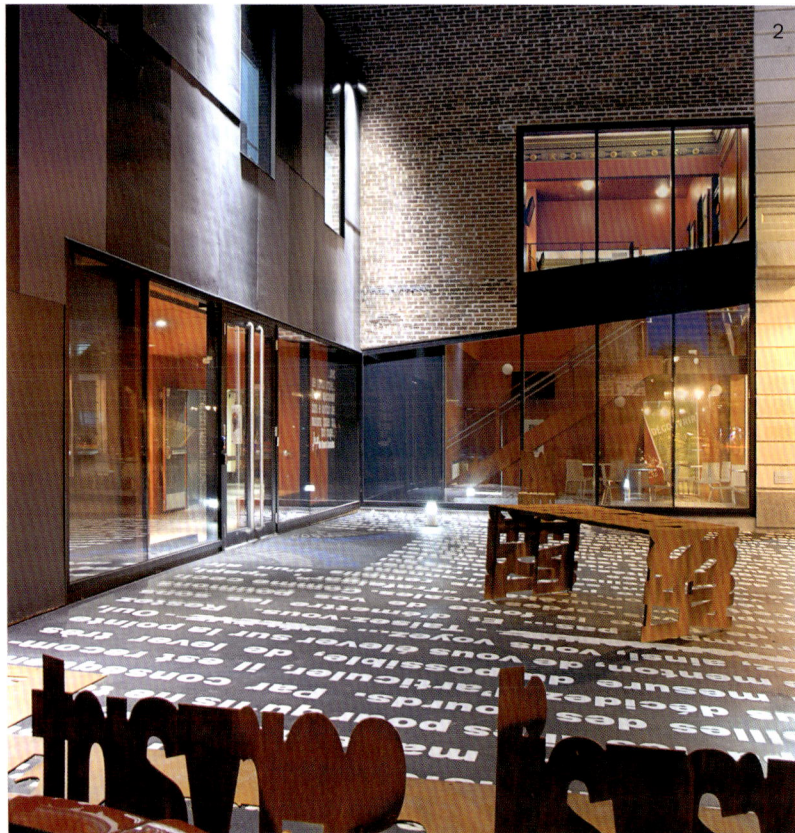

1, 2. Entrance plaza
3. Entrance plaza detail
4. Entrance lobby & ticketing

1、2. 入口广场
3. 入口广场细部
4. 入口大厅和售票处

1. Foyer
2. Bureau
3. Fred Barry Hall
4. Corridor
5. Release of scene
6. Scene
7. Denise Pelletier Hall
8. Bar
9. Vestibule
10. Ticketing

1. 门厅
2. 办公室
3. 弗莱德·巴里厅
4. 走廊
5. 布景装卸区
6. 布景区
7. 丹尼斯·佩雷蒂尔厅
8. 酒吧
9. 门廊
10. 售票处

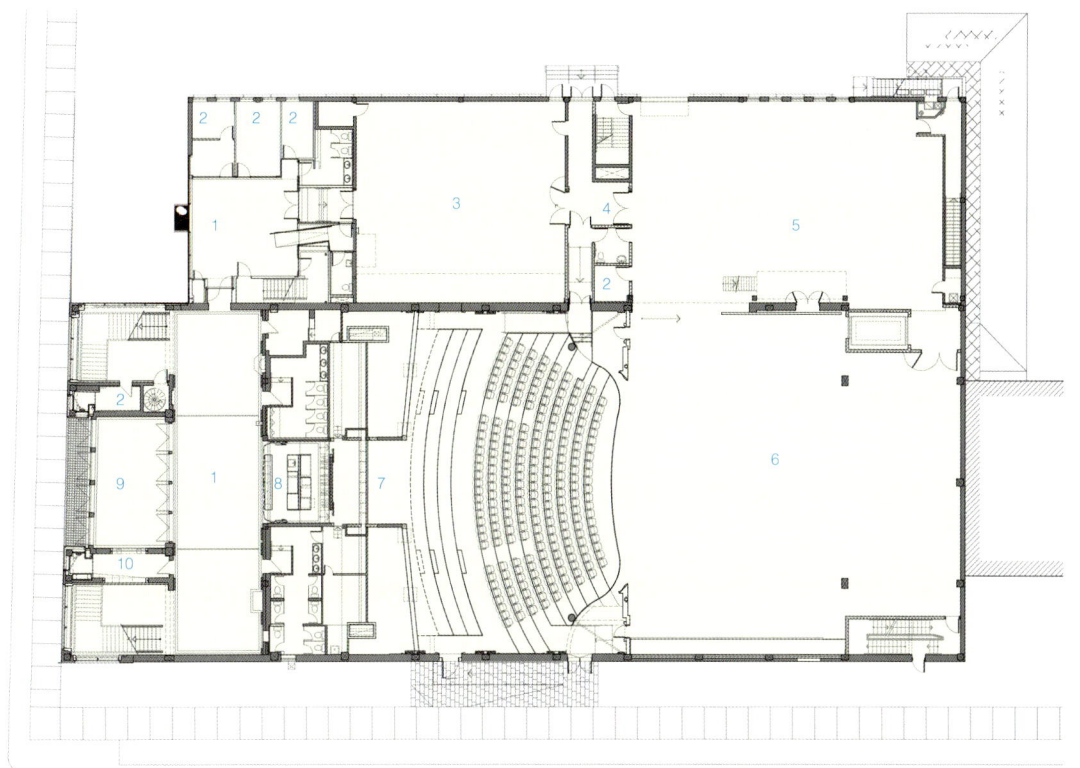

1

1. Foyer/resting
2. View show entrance plaza from foyer
3. Foyer and window-view

1. 门厅/休息区
2. 从门厅看入口广场
3. 门厅和窗外风景

1. Overview of Denise Pelletier Hall
2. Seating of Denise Pelletier Hall
3. Wall detail of Denise Pelletier Hall
4. Pass-way of Denise Pelletier Hall

1. 丹尼斯·佩雷蒂尔厅全景
2. 丹尼斯·佩雷蒂尔厅观众席
3. 丹尼斯·佩雷蒂尔厅墙面细部
4. 丹尼斯·佩雷蒂尔厅走道

贝尔格莱德剧院 # Belgrade Theatre

Location: Coventry, United Kingdom **Designer:** Stanton Williams **Completion date:** 2007 **Photos©:** Morley von Sternberg, Stanton Williams **Site area:** 2,192 square metres (new built) **Awards:** Nominated for the 2010 Federal Republic of Germany Award/RIBA National Award, 2008/RIBA West Midlands Award, 2008/D&AD Award, 2008/Finalist for the Civic Trust Award, 2008

项目地点：英国，考文垂 设计师：斯坦顿·威廉姆斯 完成时间：2007年 摄影师：莫里·凡·斯登伯格、斯坦顿·威廉姆斯 占地面积：2,192平方米（新建）所获奖项：2010年德意志联邦共和国奖提名；2008年英国皇家建筑师协会奖；2008年英国皇家建筑师协会米德兰兹分会奖；2008年D&AD奖；2008年市民信托奖决赛入围作品

When it was opened in 1958, Coventry's Belgrade Theatre was the first all-new professional theatre to be built in Britain for twenty years, and the country's first purpose-built civic theatre. Both the building and the artistic programme represented a new age. By the turn of the twenty-first century, however, the theatre's facilities were proving inadequate. The designers were commissioned in 2002 to provide the Belgrade Theatre with a second auditorium, expanded foyers, and improved backstage facilities.

The form of the new building responds to the jumps in scale that exist in the surrounding urban fabric, and also anticipates and takes on the future high-density development that will form its new setting. It is thus both bold and dynamic. Architecturally, the spaces that it embraces, and that it implies around itself, are as important as the form itself. The building pushes upwards to establish vertical space and to assert the theatre in the anticipated new context. In a series of stepped moves, it locks back into the existing building. The cubic forms orchestrate the context and then fold in to the interior volumes. Light, movement and energy slide inside and outside between the volumes.

The final form of the building was developed through a sculptural process. In essence it is derived from a main cube (accommodating the B2 studio theatre) and a sub-cube (accommodating the rehearsal room). These are dislocated from each other to create circulation spaces and a terrace. This form is then developed in the most direct way possible – with simple (and economical) materials and colours that articulate the pieces. The B2 studio theatre is housed within a concrete cube. The unusually high volume gives proportions that "draw the walls in" to increase the sense of intimacy within the space. In a way the designers have created a "found space" and this then allows a wide range of theatrical propositions and dramatic configurations to inhabit the space now and into the future.

The original design was developed as a timber horseshoe-shaped courtyard theatre (the Shakespearean "wooden O"). This design was eventually abandoned in favour of a non-specific orthogonal form that would give a greater degree of performance flexibility. The final design is thus a series of suspended steel gantries with three balcony levels one of which is a technical gallery.

考文垂的贝尔格莱德剧院创立于1958年，是英国那20年来第一座全新的专业剧院，也是英国最早的特定公民剧院。剧院的建筑形式和艺术剧目都标志着一个新时代。在21世纪的转折点，剧院设施急需改造。2002年，设计师受委托为剧院打造一个新的礼堂、扩建大厅、并且改善后台设施。

新建筑的造型与周边的城市环境相匹配，同时也为未来高密度的开放工程做了准备。建筑形式既大胆又动感。从建筑学上讲，建筑所包围的空间以及它暗中包围的空间要比建筑造型本身更重要。建筑在垂直空间中增添了剧院结构，通过路径与原有建筑相连。立方体造型将环境结合起来，折叠而成了室内空间。光线、运动和能量在空间内外交替滑动。

建筑的最终造型通过雕塑过程得以实现，由主体（B2工作室剧院）和附体（排练室）组成。二者互不相连，形成了流通路径和平台。这一造型直接而有效，材料经济简单，色彩鲜明。B2工作室剧院位于一个混凝土方块里。超高的空间向内拉近了墙壁，增添了空间的私密感。设计师打造的"创立空间"保证了现在和未来剧目和舞台布景的多样性。

最初的设想是打造一个马蹄铁形状的剧院（即莎士比亚的"木O"）。这一设计后来被不明确的直角造型所取代，增加了演出的灵活性。最终的设计由一系列悬起的钢铁吊架和三层包厢组成，其中一层是工艺画廊。

1,2. Facade detail
1、2. 外立面细部
3. View into new foyer at night
3. 新大厅夜景

1. New entrance and facade view
2. Foyer corridor to new auditorium

1. 新入口及外立面
2. 大厅走廊通向新剧场

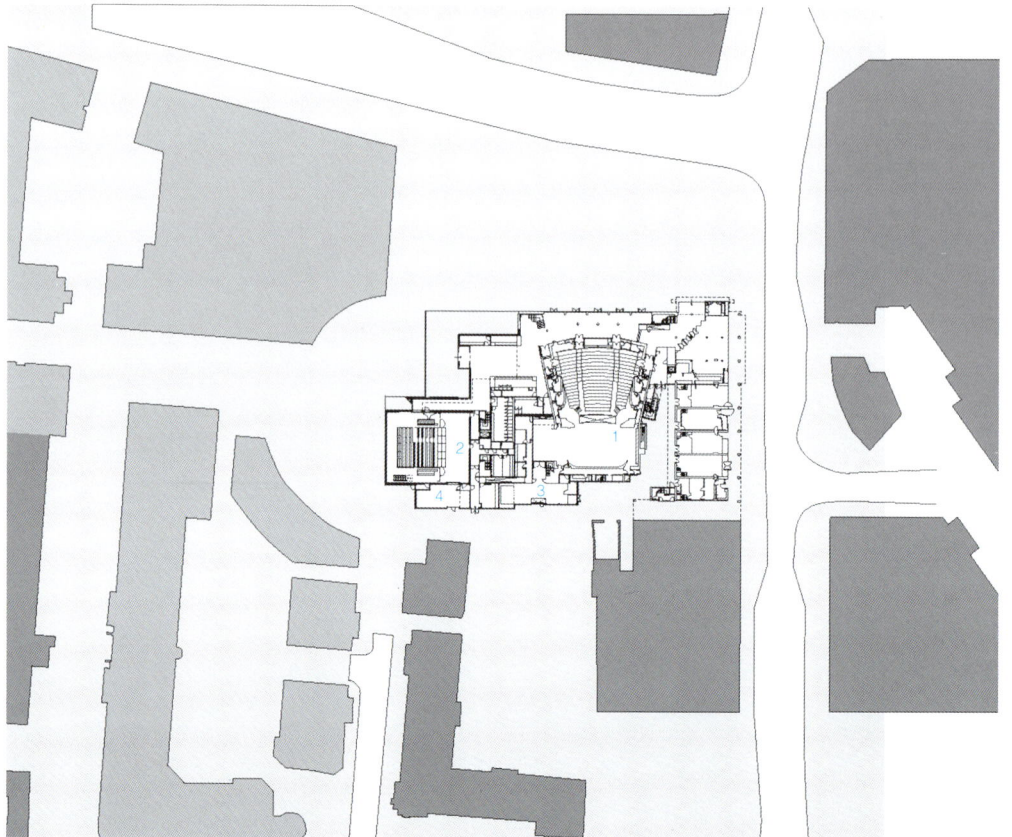

1. Existing auditorium
2. New auditorium
3. Foyer in front of existing auditorium
4. Foyer in front of new auditorium

1. 原有剧场
2. 新剧场
3. 原有剧场前的大厅
4. 新剧场前的大厅

2

1. New foyer and drink bar
2. New auditorium

1. 新大厅及饮品吧
2. 新剧场

2

动物园剧场 **Zoo Zoo**

Location: Gyeonggi-do, South Korea **Designer:** Hyunjoon Yoo Architects **Completion date:** 2010 **Photos©:** Seunghoon Yum **Area:** Lion Theatre: 379.78 square metres, Crocodile Theatre: 417.17 square metres

项目地点：韩国，京畿道 设计师：柳贤俊积攒后事务所 完成时间：2010年 摄影师：廉胜勋 面积：狮了剧场：379.78平方米，鳄鱼剧场：417.17平方米

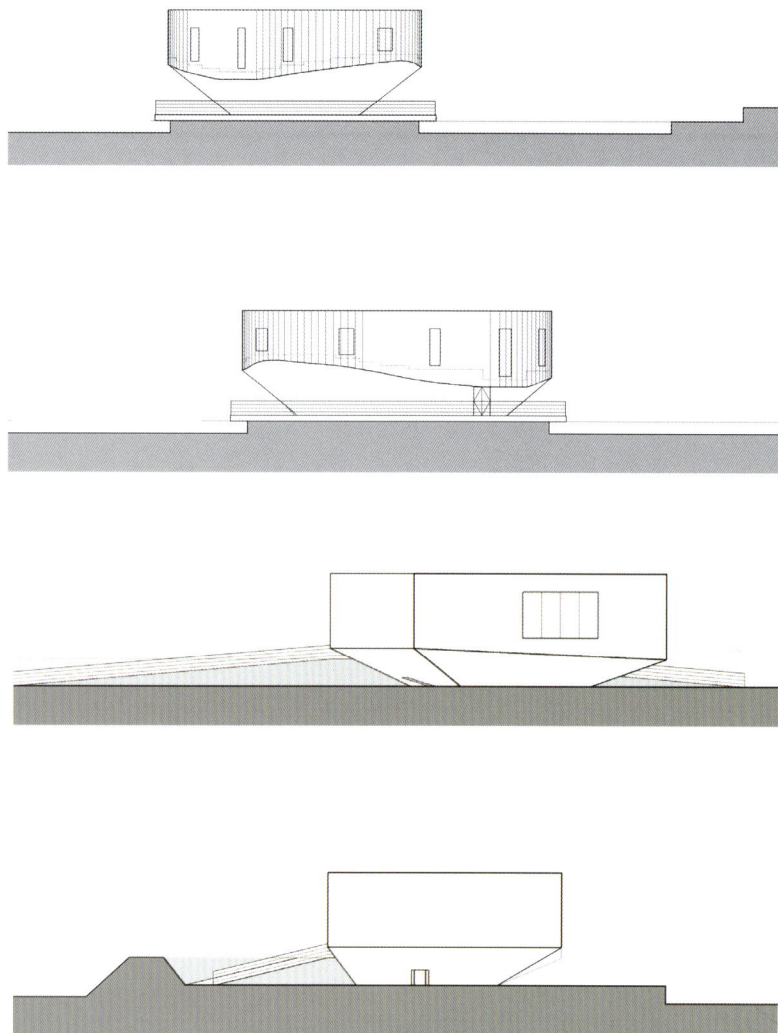

"Zoo Zoo" is filled with too much content of reptiles' exhibition hall, crocodile theatre, lion theatre, sea lion theatre, etc. within a relatively small size site. It might be required four times of current size to accommodate all kinds of programmes on the zoo. In spite of this limited condition, this zoo is always crowded with elementary students in Seoul and Geonggi-do area visit this zoo to have an experience of touching animals. The site is located near by the river and the site has a very high chance of flood during the raining season due to the low site level.

The client wanted to have lot of eve space for the visitors to avoid shower and direct sunlight for lunch time. The best solution for three conditions: maximising the use of small site, avoiding flood and avoiding rainfall was simple. The answer was "piloti". However piloti can cause too high building blocking the beautiful sky and scenery. So the designers had to design low height building with piloti space by utilising underneath of sloped seats of theatre.

There are two buildings. One is a "lion theatre" and the other is a "crocodile theatre." While keeping the design concept of utilising underneath of the seats, the building shape is defined by the shape of the site. The lion theatre mass is molded by the parallelogram shape site and 20 degree seats slope of the client's request. The Crocodile Theatre mass is molded by the distorted round shape island site and 37 degree slope seats. This condition produced three dimensional curve line edges at the elevation. This arbitrary looks alike mass are actually cast logically by the requested seat slope and site shape.

Current Zoo Zoo architectural environment is chaotic because of complex programme, too many stone sculptures of the client's collection and many kinds of trees. Therefore newly constructed building has to be simple to give a visual rest, and to work as a landmark giving visitors sense of relative location within the zoo. As a solution for these needs, architects designed the building minimal white mass of asymmetry shape.

动物园由爬行展厅、鳄鱼剧场、狮子剧场、海狮剧场等组成，所在地点相对较小。为了满足全部项目的空间需求，动物园将现在的规模扩大了四倍。除了场地限制之外，动物园里常有来自首尔和京畿道地区的小学生，他们到动物园来亲近公务。剧场位于河边，地势较低，在雨季很容易洪水泛滥。

委托人希望为游客提供避雨的空间，同时保证午餐时光温暖的阳光。最佳的解决方法是：最大化空间配置、避免洪水和雨水。这些都很简单，通过底层架空柱就可以实现。但是底层架空柱会让建筑过高，看不到美丽的天空和风景。因此，设计师必须通过剧院内的倾斜座椅设计一个相对较矮的底层架空柱。

一共有两座建筑，分别为"狮子剧场"和"鳄鱼剧场"。二者都运用了座椅底座的设计理念，但是建筑外形随着场地的形状不同而不同。狮子剧场呈平行四边形，座椅坡度20度。鳄鱼剧场呈扭曲的圆形，座椅坡度37度，形成了三维曲线边缘。事实上，这一随意的造型是由座椅的坡度和场地的形状所决定的。

目前，动物园的建筑环境十分混乱——复杂的项目、过多的石雕、过多的树种。因此，新建的建筑必须简单，让视觉清新，同时也要起到路标的作用，引领游客寻找动物园内的景点。为了满足这些需求，建筑师将建筑设计成简约的白色几何造型。

1. General view　　1. 建筑全景
2. Crocodile Theatre　　2. 鳄鱼剧场

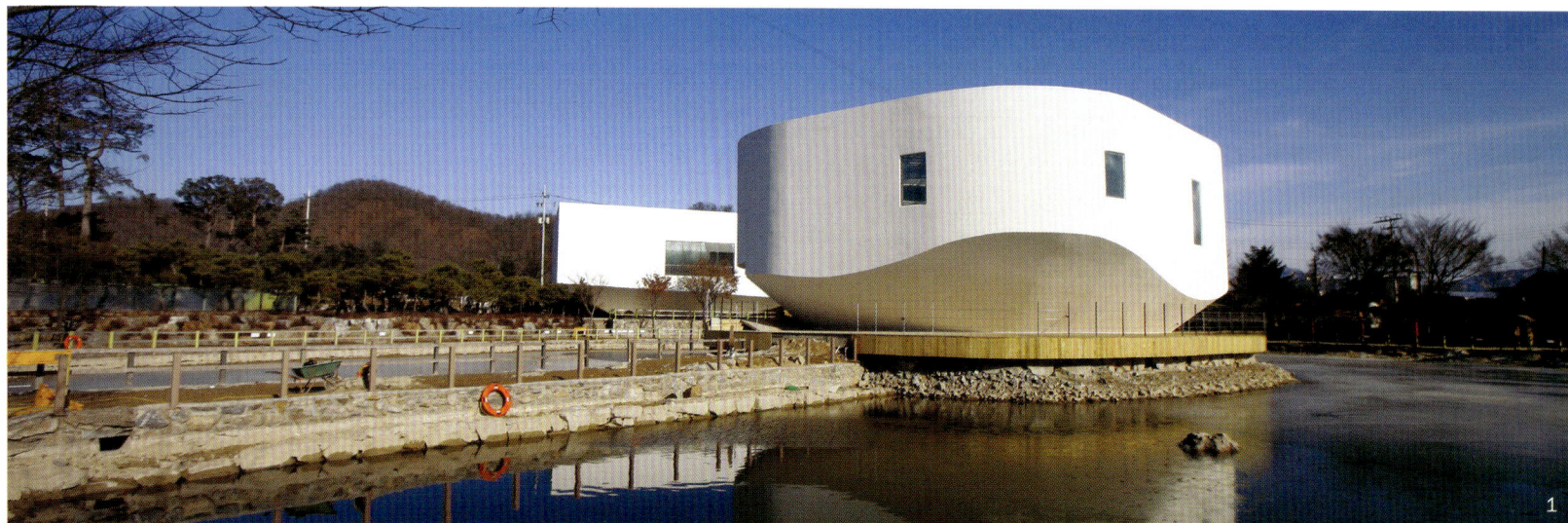

1

2

3

1. Lion Theatre
2. Façade detail of Lion Theatre
3. Distant view of Lion Theatre
4. Interior of Crocodile Theatre

1. 狮子剧场
2. 狮子剧场外观细部
3. 狮子剧场远景
4. 鳄鱼剧场内部

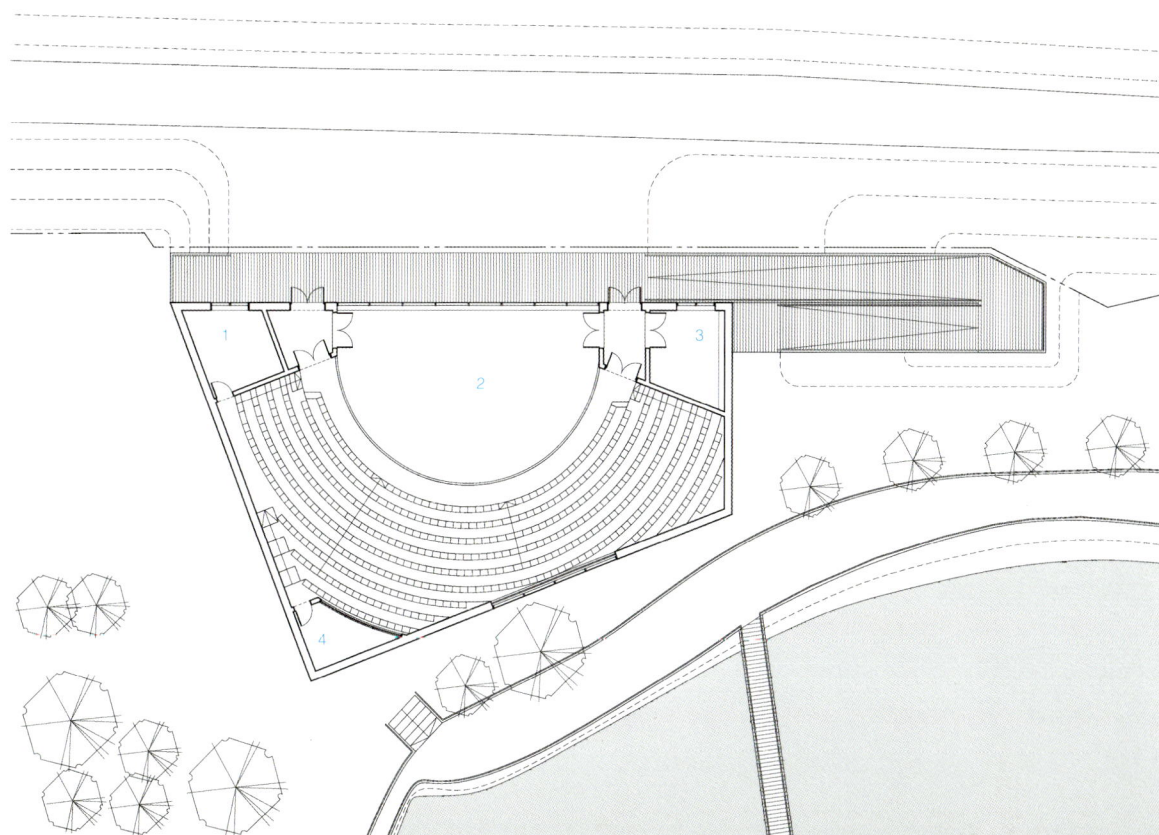

1. Sub room
2. Learning room
3. Gallery
4. Studio

1. 辅助室
2. 学习室
3. 走廊
4. 工作室

比吉尔莫尔公园剧院 **Bijlmer Park Theatre**

Location: Amsterdam, The Netherlands **Designer:** Paul de Ruiter **Completion date:** 2009 **Photos©:** Pieter Kers **Gross floor area:** 2,010 square metres

项目地点：荷兰，阿姆斯特丹 设计师：保罗·德·瑞特 完成时间：2009年 摄影师：皮特·卡尔斯 建筑面积：2,010 平方米

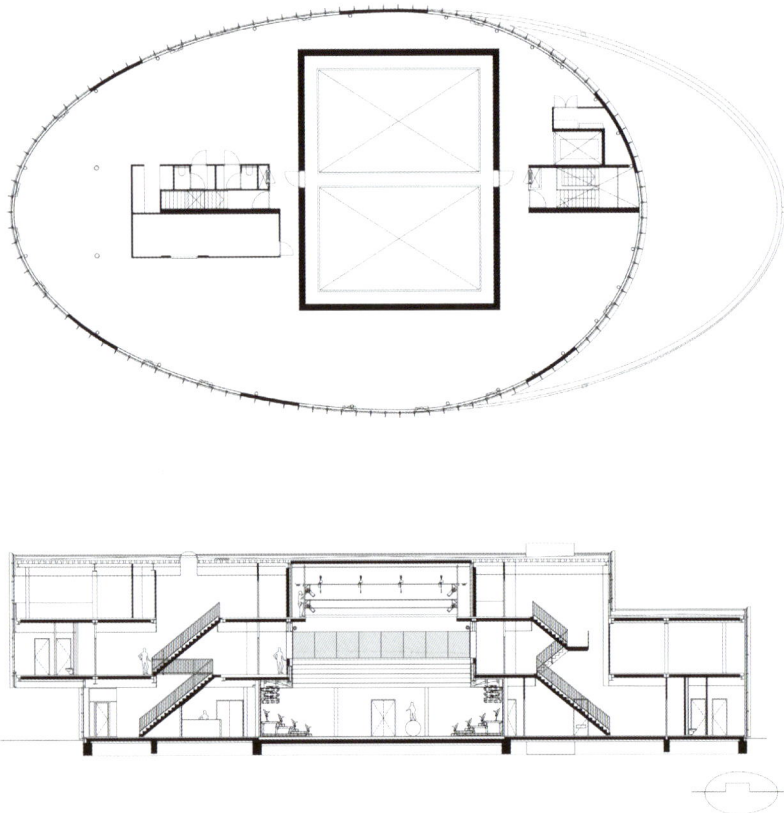

The site of the building is specified in the urban development plan. The building is located in the heart of the Bijlmer neighbourhood at the edge of the Bijlmer Park, beside the lake. It is public and accessible, and its position beside the water gives extra dynamism to this image due to the reflections in the water.

The building consists of an ellipse shape, with the upper two floors slightly displaced in relation to the ground floor. This provides a covered entrance area located in a logical position in the urban development plan's routing. The elliptical shape of the building did mean that it was necessary to search for a financially viable way of reproducing this rounded shape in the partially glass façade. The solution was found in a combination of wooden slats and vertical aluminium strips placed against the steel and glass sections of the façade. This means that the intersection points of the segmented façade are not visible and the building has a rounded, dynamic and somewhat abstract appearance that changes continually as you walk around it.

During the day, the striking shape of the building makes it clearly recognisable, while it is conspicuous in the evening because of its colour, which can be altered to fit the occasion. This is made possible by the use of LED lighting. A line of light is fitted behind the steel façade, shining downwards. Because this light shines against the steel façade and the wooden slats, the building acquires an appearance of transparency, as if the light is coming from inside the building. The illumination of the building increases the level of safety and makes the building clearly visible from the urban surroundings.

One requirement that was specifically identified during the workshops was the need for daylight in the main auditorium. Lessons and rehearsals would take place here during the day, and a good level of daylight access is very important for the atmosphere and sense of orientation. For this reason, a glass surround was created on the ground floor all around the main auditorium. This solution not only allows a maximum capture of light, it also makes it possible for parents and others who may be interested to watch lessons and rehearsals unobtrusively. These windows can be darkened to keep out the light when performances are held.

In addition to the main auditorium, the building has a spacious foyer, rehearsal rooms, three studios, storage rooms, dressing rooms, a sewing room, meeting facilities and offices. The building accommodates the four user groups for dance class of the Amsterdam School of Arts. For all these users the three storeys are arranged. The main auditorium extends to the height of all three storeys and one of the studios is two storeys high. On the top storey, the bridges for the operation of lighting and set management are integrated into the concrete floor. This is a practical and inexpensive solution that makes the bridges safe and easily accessible.

这座文化建筑是城市发展规划的一部分，位于比吉尔莫尔公园边缘的社区中央位置，就在湖畔。建筑极具公共性和可入性，湖水的倒影让建筑充满了活力。

剧院呈椭圆形，上两层楼与一楼的空间轻微错开。建筑的带顶入口处在城市发展路径的有利位置之上。建筑的椭圆造型要求其设计找到一个经济可行的方案来重建玻璃外墙。木板条和垂直铝条的结合抵住了钢铁和玻璃外墙。这意味着分割式外墙交叉点不在可见，建筑的圆形造型动感而抽象，并且随着人们的走动而持续变换。

1, 2. Main facade   1, 2. 建筑外观
3. Segmented facade   3. 分割式外墙

1

2

3

白天，建筑的造型十分抢眼；夜晚，LED灯光所营造的变换的色彩也令人惊叹。钢铁外墙上嵌入了一道光线，照亮了下方。由于灯光以钢铁表面和木板条为背景，僵住看起来十分透明，仿佛灯光是从建筑内部射出的一样。建筑的灯光增添了安全性，也让建筑在城市环境中脱颖而出。

剧院最重要的设计要求之一是保证主礼堂内的自然采光。白天，这里将进行授课和排练，自然采光对气氛和方向感十分重要。因此，主礼堂的外围是玻璃制成的。这不仅保证了采光的最大化，还让家长和其他对教学、排练感兴趣的人可以自由围观。表演时，这些窗户会被遮住。

除了主礼堂之外，剧院还拥有一个宽敞的大厅、若干排练室、三间工作室、仓库、更衣室、缝纫室、会议和办公设施。建筑为阿姆斯特丹艺术学院的四个用户群提供舞蹈教室。主礼堂有三层楼高，其中的意见工作室也有两层楼高。顶楼，灯光操作和背景管理天桥与混凝土地面结合在一起。这一解决方案实用而经济，让天桥安全而容易进入。

1. Exterior night view
2. Entrance night view

1. 外观夜景
2. 入口夜景

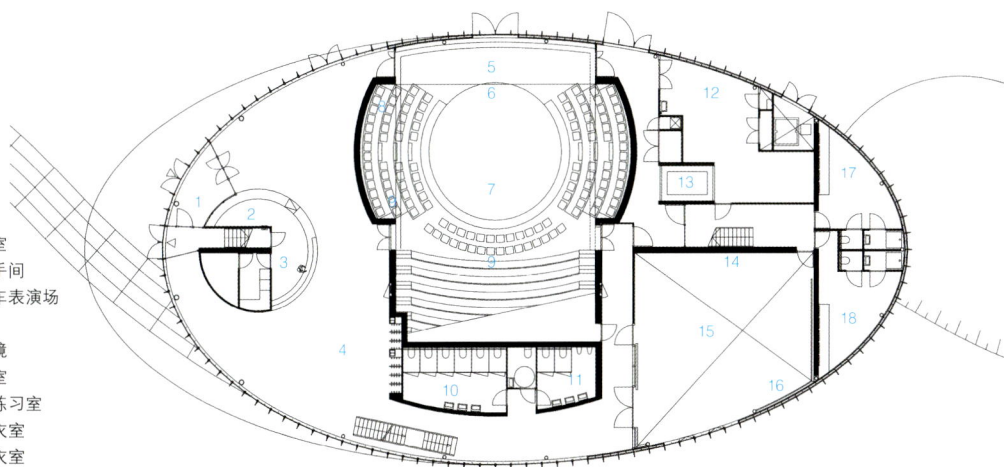

1. Entrance	10. Powder room(women)	1. 入口	10. 电力室
2. Reception	11. Toilet room(men)	2. 前台	11. 男洗手间
3. Bar	12. Bicycle	3. 吧台	12. 脚踏车表演场
4. Foyer	13. Lift	4. 门厅	13. 电梯
5. Back stage	14. Wall mirror	5. 后台	14. 墙面镜
6. Circus arena	15. Studio	6. 马戏表演场	15. 工作室
7. Great hall	16. Ballet-bars	7. 大表演厅	16. 芭蕾练习室
8. Lifting track	17. Girls' locker room	8. 升降轨道	17. 女更衣室
9. Seats at banks	18. Boys' locker room	9. 观众席	18. 男更衣室

1

1. Spacious hall & concrete floor
2. Hallway
3. Foyer
4. Stairway

1. 宽敞的大厅和水泥地面
2. 走廊
3. 门厅
4. 楼梯

2

3

4

1. Rehearsal room
2. Main auditorium
3. View show auditorium through glass window

1. 排练室
2. 主礼堂
3. 透过玻璃窗看礼堂

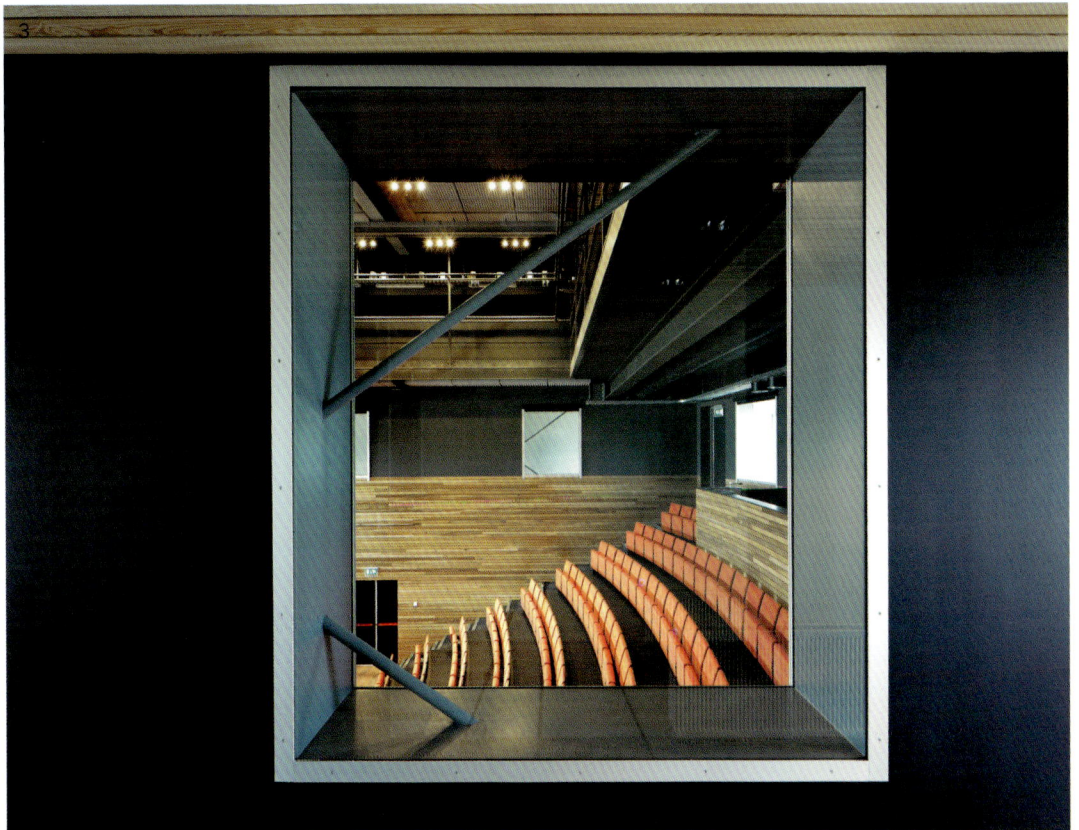

大运河广场剧院和商业开发项目

Grand Canal Square Theatre and Commercial Development

Location: Dublin, Ireland **Designer:** Stefan Blach/Studio Daniel Libeskind **Completion:** 2010 **Photos©:** Studio Daniel Libeskind/Ros Kavanagh/Jarek Matla Photography **Building area:** 68,180 square metres

项目地点：爱尔兰，都柏林 设计师：斯蒂芬・布拉齐/丹尼尔・李伯金斯工作室 完成时间：2010年 摄影师：丹尼尔・李伯金斯工作室、罗斯・卡瓦纳、加瑞克・迈特拉摄影 建筑面积：68,180平方米

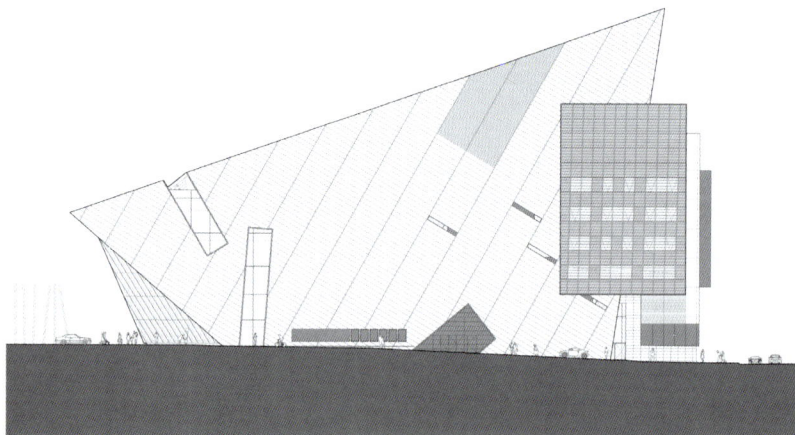

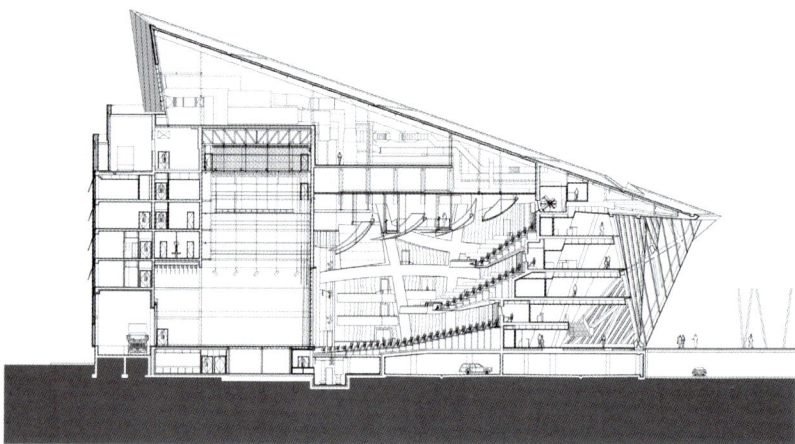

The concept of the Grand Canal Square Theatre and Commercial Development is to build a powerful cultural presence expressed in dynamic volumes sculpted to project a fluid and transparent public dialogue with the cultural, commercial and residential surroundings whilst communicating the various inner forces intrinsic to the Theatre and office buildings. This composition creates a dynamic urban gathering place and icon mirroring the joy and drama emblematic of Dublin itself.

The 2,000-seat Grand Canal Theatre is a landmark that creates a focus for its urban context, specifically Grand Canal Square, the new urban piazza at the waterfront of Grand Canal Harbour. The architectural concept of the Theatre is based on stages: the stage of the Theatre itself, the stage of the piazza, and the stage of the multiple level Theatre lobby above the piazza. The Theatre becomes the main façade of a large public piazza that has a five-star hotel and residences on one side and an office building on the other. The piazza acts as a grand outdoor lobby for the Theatre, itself becoming a stage for civic gathering with the dramatic Theatre elevation as a backdrop offering platforms for viewing. From its rooftop terrace, the Theatre offers spectacular views out over the Dublin Harbour.

The Theatre is integrated into the Commercial Development by office buildings that include 45,500 square metres of leasable office and retail space. With their twin façades, glazed atriums and landscaped roofs, the two office blocks offer sustainable state of the art work environments. By designing multi-storey glazed atriums, the commercial buildings integrate with the adjacent retail, residential, cultural and public space components. Three prominent entrances make the buildings accessible from Grand Canal Square, Misery Hill and from Cardiff Lane. Although both offices are designed in the same architectural language, each responds to its site uniquely. Two Grand Canal Square (south block), which is adjacent to the new 2,000-seat theatre, opens up towards the Square, while Four & Five Grand Canal Square (north block), in conjunction with the Theatre, form a dramatic gateway to Dublin Harbour.

大运河广场剧院和商业开发项目旨在打造一个强有力的动态文化空间，与周边的文化、商业和住宅形成流畅而透明的公共对话，向剧院和办公楼传达各种内在的力量。这是一个动感的城市集会空间，也是都柏林戏剧和快乐的象征。

拥有2,000个坐席的大运河剧院是当地的标志性建筑，特别引入注目的是位于港口的新建的大运河广场。剧院的建筑理念以舞台为基础：剧院的舞台、广场的舞台、剧院大厅的舞台。剧院是大型公共广场的正面，广场的一侧是五星级酒店和住宅，另一侧是办公楼。广场是剧院的户外大厅，是市民们集会的重要舞台。从剧院的屋顶平台可以俯瞰都柏林港口的美景。

剧院通过办公楼与商业开发项目结合在一起。办公楼拥有45,500平方米的出租写字间和零售空间。同样的外墙、玻璃中庭和景观屋顶让两座办公楼为艺术工作环境提供了可持续的空间保证。多层的玻璃中庭让商业楼集零售、住宅、文化和公共空间于一身。三个入口分别通着大运河广场、米泽瑞山和加迪夫路。尽管两幢办公楼采用了同样的设计理念，它们的设计随着场地的变化而各不相同。二号大运河广场（南）邻近剧院，朝向广场；四号和五号大运河广场（北）连着剧院，朝向都柏林港口。

1. General view   1. 建筑全景
2. Office building façade   2. 办公楼外观

1. Plaza
2. Façade
3. View from the street
4. Office building façade

1.广场
2. 建筑外观
3. 从街道看建筑
4. 办公楼外观

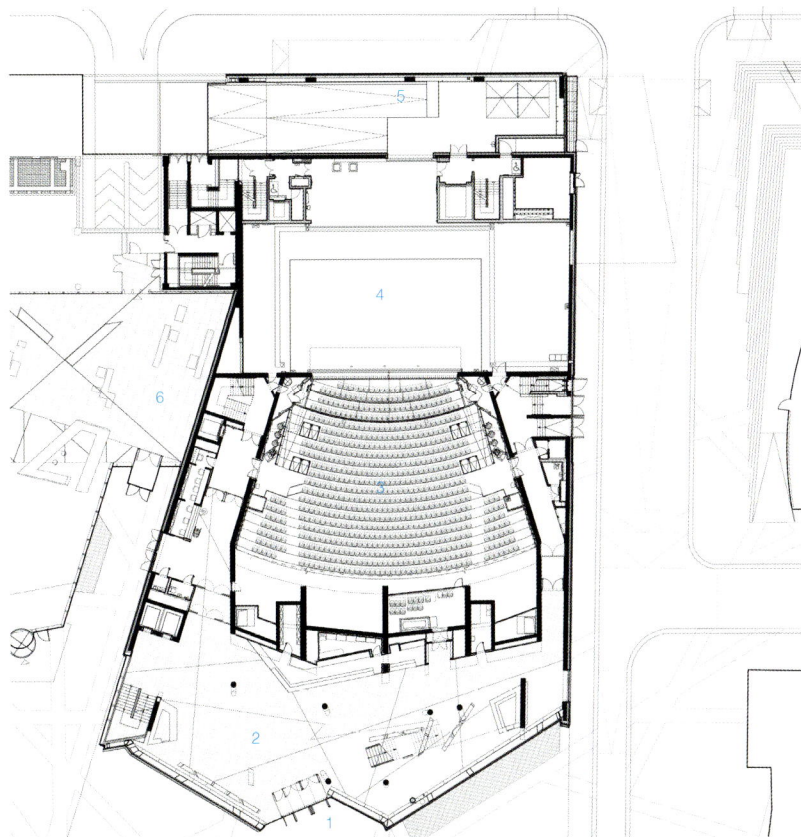

1. Theatre entrance
2. Main foyer
3. Stalls
4. Stage
5. Loading bay
6. South block commercial building

1. 剧院入口
2. 门厅
3. 正厅前座
4. 舞台
5. 装卸区
6. 南侧商业建筑

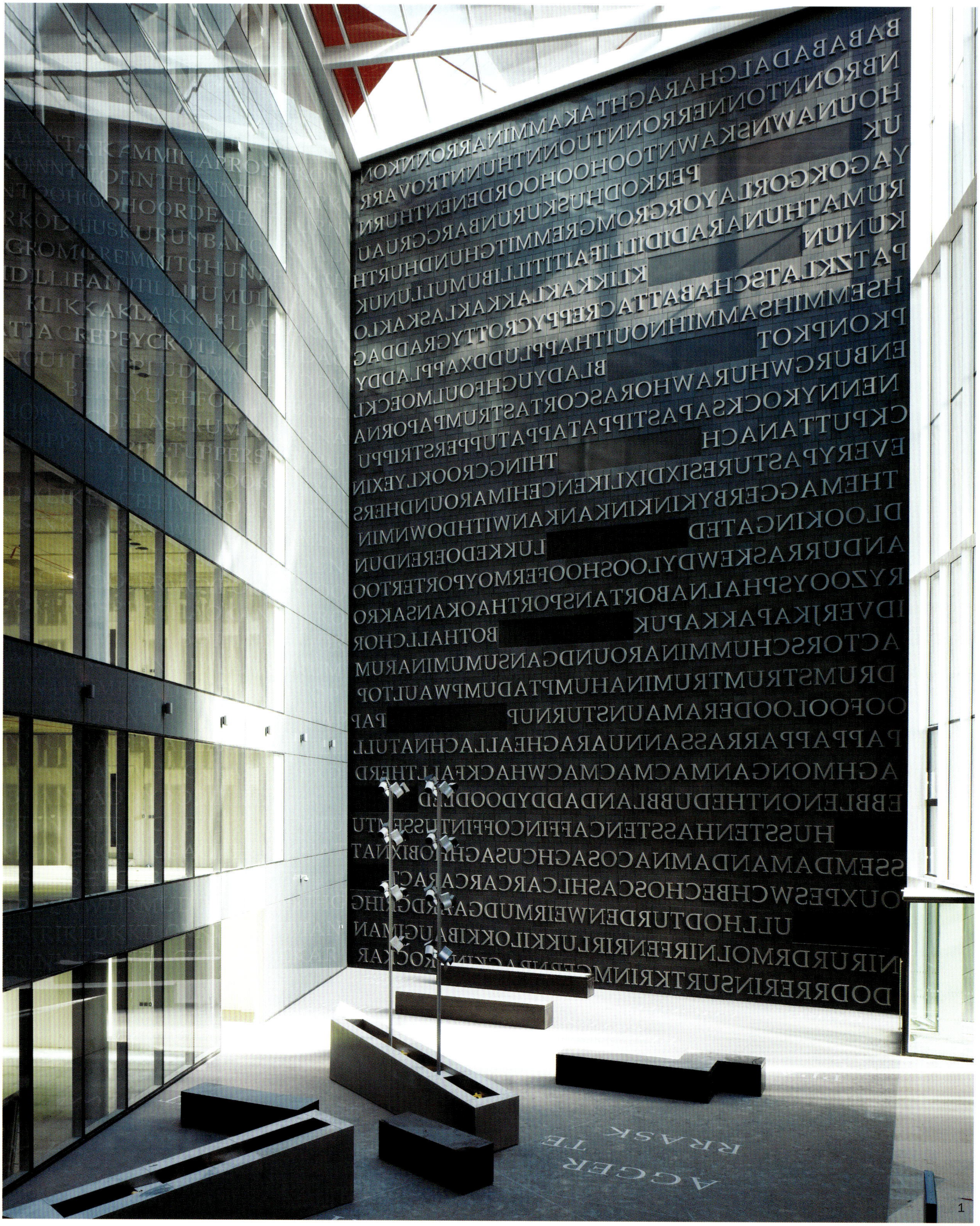

1

1. Office building plaza
2. Entrance lobby
3. Theatre hall, photo© Ros Kavanagh

1. 办公楼广场
2. 入口大厅
3. 剧院大厅© Ros Kavanagh

1. Opening night interior
2, 3. Auditorium

1. 首演之夜
2、3. 观众席

蒙特勒伊剧院 # CDN Montreuil

Location: Paris, France **Designer:** Dominique Coulon, Steve Letho Duclos, Sarah Brebbia, Arnaud Eloudyi, Olivier Nicollas **Completion date:** 2008 **Photos©:** Jean Marie MONTHIERS **Surface area:** 2,600 square metres

项目地点：法国，巴黎 设计师：多米尼克·库隆、史蒂夫·勒多·杜卡拉斯、莎拉·布莱比亚、阿诺德·埃罗德伊、奥利维尔·尼古拉斯 完成时间：2008年 摄影师：简·玛丽·蒙蒂尔斯 面积：2,600平方米

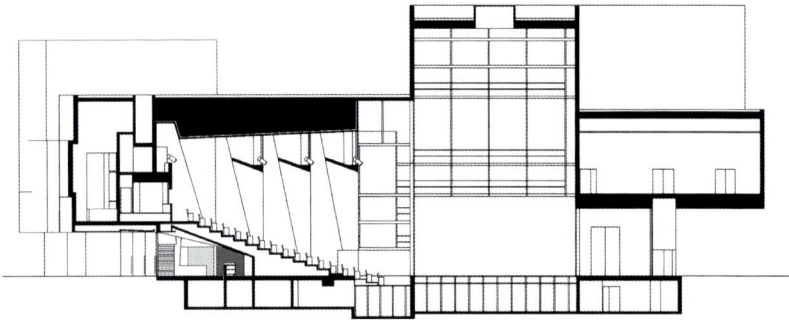

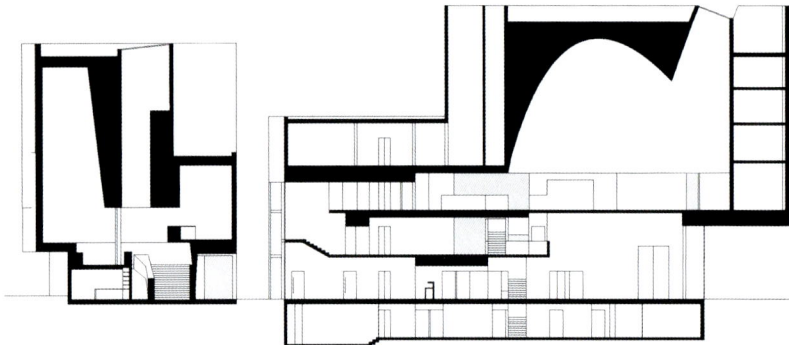

In Montreuil, located northeast of the urban centre of Paris, a master plan in recent years has been realised for the regeneration of the urban centre by Alvaro Siza. The project, with the theatre at its heart, shows the hill of the site: the buildings like the fingers of a hand reveal the slope and increase the perspectives. The theatre has no front or back. In response to the two diagonal areas adjoining the building, the auditorium pivots by 15 degrees, rises up and opens simultaneously onto the two public areas without installing a hierarchy. This rotation also resolves the problem of the town hall, the façade of which is not exactly in line with the plot – a lateral projection out of the main volume makes up for the difference, firmly positioned opposite the entrance to the monumental 1935's building.

When you walk around, the building is mysterious because of its scales variations: sometimes intimate, sometimes monumental. It looks like a sculpture. The façades are made with white concrete, made on the construction site. This white cement contains photocatalytic particles that allow it to oxidise the organic and inorganic air pollutants in the presence of air and light. This photocatalytic action gives to the construction a perpetual esthetic and more luminous surfaces.

Inside, the space is compressed but elusive. It seems to be generous because of the views. The volumes of the staircase, the entrance hall and the rehearsal area, which wrap around the auditorium, are oversized so that they are not crushed by the mass of the stage and backstage area. The result is, on the outside, an ensemble with perfect proportions and, on the inside, staggering empty spaces that can be freely restructured or even sculpted. In response to the fist-shaped mass that asserts its rule over the shapes present on the site, there is, as soon as the threshold has been crossed, a chain of sequences declining the figures of tightening and stretching. This complexity is emphasised by colour, by its presence that becomes mass, or by contrast from its absence that becomes light. The colour of the path that leads to the auditorium goes from red to black with all the nuances possible including stairways and corridor. The colours of the wood floor (padouk and wenge) change too when the visitor approaches the auditorium. The continuing movement sequences bring the spectator to other space. Slowly, it makes people ready to enter, make silence and see the show.

阿尔多·西扎为巴黎市中心东北侧的蒙特勒伊打造了一个整体规划，以复兴城市中心。项目的中心是一个剧院，仿佛张开的手指，显露了坡度，也增强了透视感。剧院不分前后，与连接建筑的两个对角线区域相适应，礼堂偏转了15度，与它们自然地连接了起来。这一转动还解决了市政厅与场地不在同一直线上的问题。市政厅设在建筑之外，正对建于1935年的历史建筑的大门。

空间和规模的变奏让建筑显得十分神秘，有时私密，有时宏大，仿佛一件雕塑。建筑表面由白色混凝土组成，里面还有光学颗粒，可以氧化空气和光线中的有机和无机空气污染物。这一光学活动让建筑表面看起来更具美感，也更光亮。

内部空间紧凑而难以捉摸。窗外的景色让它看起来十分开阔。楼梯、入口大厅和排练区环绕着礼堂，看起来十分宏大，不次于舞台和后台空间。这样一来，外部空间拥有完美的比例，内部空间也可以进行任意的重建和改造。拳头造型的建筑主体在场地中拥有压倒性的地位，旁边环绕着一系列紧扣或弯曲的手指。建筑的色彩十分引人注目，聚合起来则形成了色块，对比之下形成了光线。通往礼堂的走道的色彩从红色渐变为黑色，其中经过了楼梯和走廊。紫檀和黑铁木地板的色彩也逐渐变换。延续的运动序列将人们从一个地点带到另一个地点，逐渐让人们安静下来，做好观看演出的准备。

1. Façade detail   1. 外观细部
2. Interior structure   2. 内部结构
3. Building and trees   3. 建筑和树木
4. Main facade view   4. 建筑外观

3

4

1. Interior hallway
2. View from upper foyer

1. 内部走道
2. 从上层门厅往下看

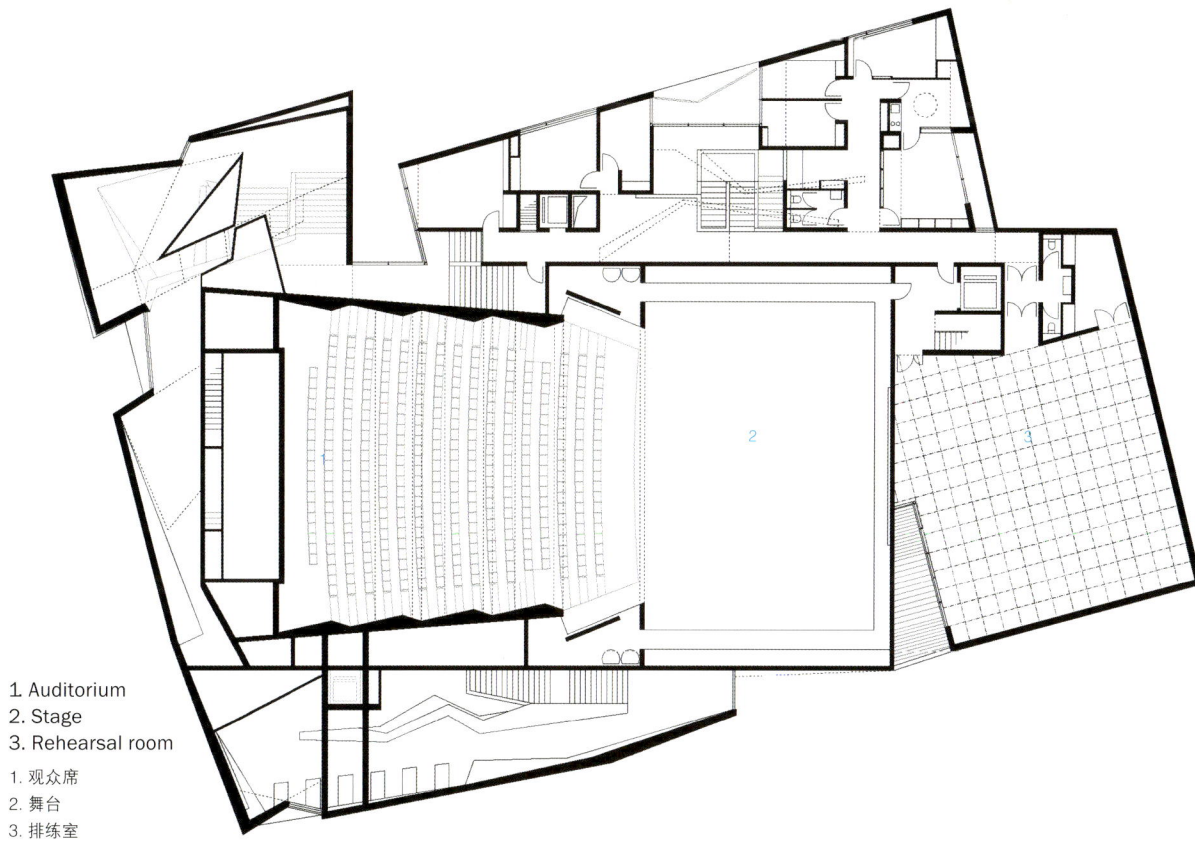

1. Auditorium
2. Stage
3. Rehearsal room

1. 观众席
2. 舞台
3. 排练室

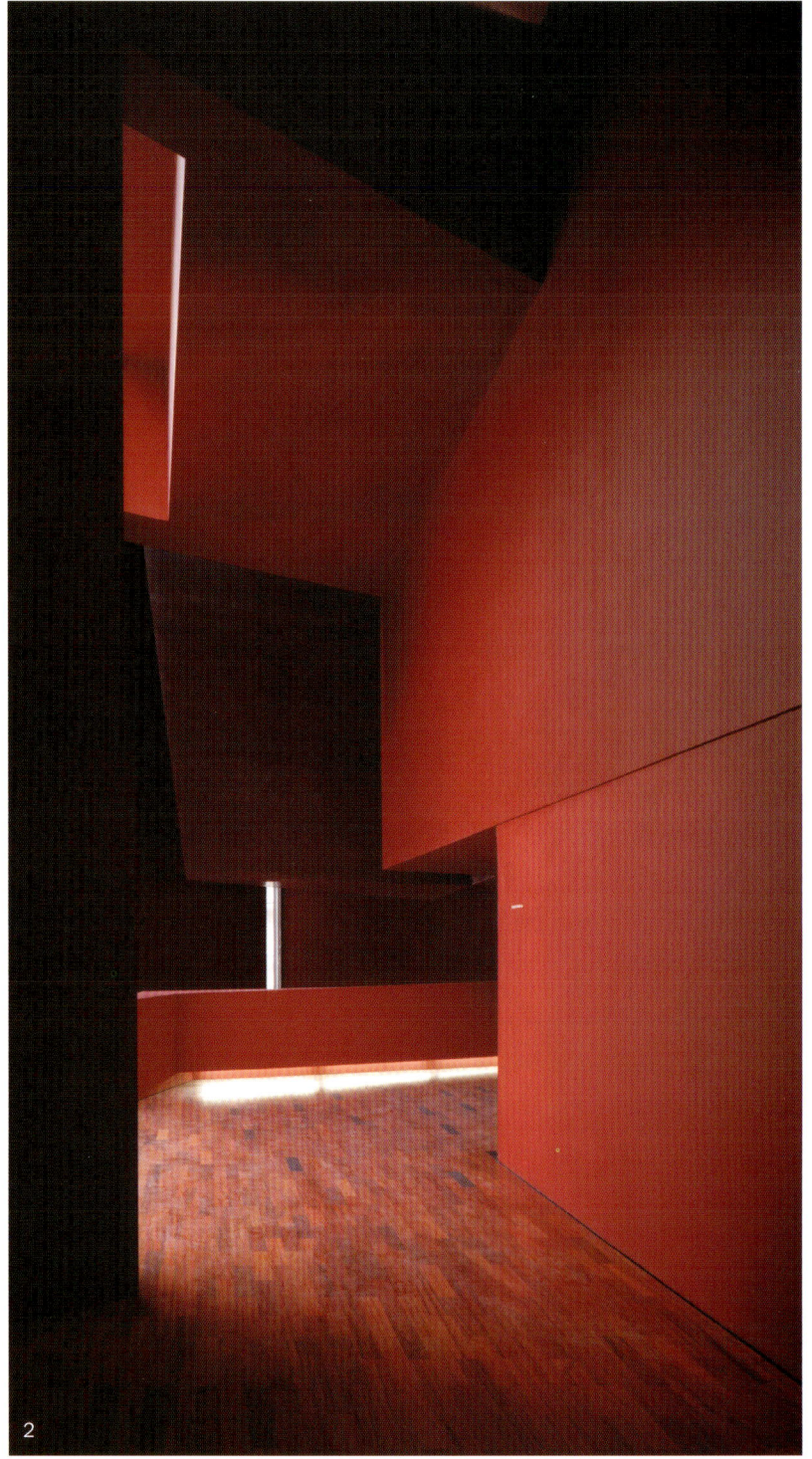

1. View from upper foyer
2. Interior hallway
3. Bar/resting area

1. 从上层门厅往下看
2. 内部走道
3. 吧台/休息区

3

1. Auditorium
2. Rehearsal room
3. View trees from the toilet

1. 观众席
2. 排练室
3. 从洗手间看外面的树木

四季表演艺术中心

Location: Toronto, Canada Designer: Diamond and Schmitt Architects Completion date: 2006
Photos©: Tom Arban, Steven Evans Awards: 2009 Royal Architectural Institute of Canada (RAIC)
Awards of Excellence - Innovation in Architecture, Honorable Mention; 2008 International Architecture
Awards, The Chicago Athenaeum: Museum of Architecture and Design; 2007 Toronto Urban Design
Awards - Public Buildings in Context, Award of Excellence
2007 BusinessWeek/Architectural Record Awards - Citation for Excellence;

项目地点：加拿大，多伦多 设计师：戴尔蒙德和施密特建筑事务所 完成时间：2006年 摄影师：汤姆·阿尔班、斯蒂芬·伊文思 所获奖项：2009年加拿大皇家建筑学院优秀创新建筑奖提名；2008年国际建筑奖，芝加哥图书馆：建筑与设计博物馆；2007年多伦多城市设计奖——公共建筑传承优秀奖；2007年《商业周刊》/《建筑实录》奖；

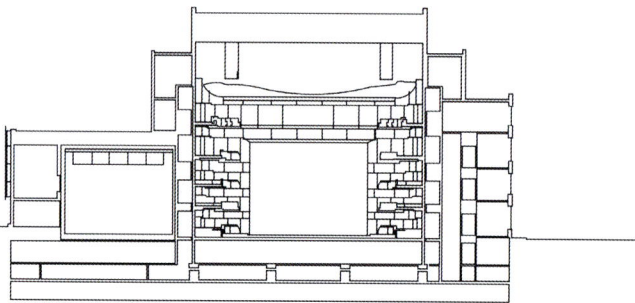

Cross-section through auditorium　礼堂剖面图

Four Seasons Centre for the Performing Arts

The award-winning Four Seasons Centre for the Performing Arts is a building in harmony with its context, designed to embrace and engage the city around it. An understated composition of rectangular shapes, the opera house's exterior is reflective of the orthogonal nature of the city's street grid contrasting with the curvilinear form of the auditorium within. Entry into the Four Seasons Centre is from a plaza at the north-west corner of the site, allowing the bulk of the building to define the street edge rather than be set back.

Conceived as a lantern in the cityscape, the City Room encloses what can be seen as an extension of the sidewalk and public realm providing passersby glimpses of an animated interior. By revealing the activity associated with attending a performance, a new audience is cultivated and a new relationship formed between the patron and the city.

An amphitheatre within the City Room offers spectacular views of the city and is programmed with over 100 free lunchtime and early evening performances that play to over 20,000 people annually. This multi-purpose space allows the opera company to fulfill its mandate to create an inviting opera house that has animation and engagement with the city outside of the regular evening and weekend performance times. The sight of audiences two storeys up enjoying a wide spectrum of performed art, in contrast to the bustle on University Avenue, is now a part of the rhythm of the city.

The transparency of the public spaces is in dramatic contrast to the enclosing envelope of the auditorium, where the external world is excluded in favour of a focus on the performance within. The design of R. Fraser Elliot Hall, the five-tiered, 2,000-seat, European horseshoe-shaped auditorium at the heart of the opera house, was driven by three factors: acoustics, intimacy, and elegance. In opera, clarity of text and natural vocal tone are important, but so are warmth and resonance of the orchestral sound. To achieve the best acoustic results all audible background noise needed to be eliminated – a significant challenge as the site is adjacent to busy subway and streetcar lines. Consequently, the auditorium, stages, and rehearsal hall are designed to be an entirely separate and isolated structure within the building. Encased in a double layer of concrete and resting on nearly 500 high performance rubber pads, this chamber is built to the extremely demanding N-1 acoustic criteria and is the first structurally isolated performance facility in Canada.

四季表演艺术中心与周边环境和谐地融合在一起，融入了城市环境。低调的矩形造型反映了剧院对城市网格的敬意，也与内部的圆形礼堂形成了鲜明对比。人们可以通过场地西北角的广场进入四季中心。这样一来，建筑便可以紧邻街道边缘，而不用向后退。

城市空间被设计为城市风景的明灯，行人可以透过玻璃窗看到其活跃的内部空间。通过将演出活动展出来，演艺中心会得到更多的观众，也在赞助人和城市之间建立新联系。

城市空间里的圆形剧场可以遥望城市的美景，每年可以提供100次免费午餐和早场夜间表演，吸引20,000名观众。这个多功能空间让剧院公司打造了一个活跃而与城市紧密相连的剧院，使其在常规的夜间和周末演出之外还能与城市有更多的联系。二楼的观众可以享受到各种表演，与喧嚣的大学大道不同，使城市充满了动感的旋律。

透明的公共空间与封闭的礼堂形成了鲜明对比。礼堂脱离了外部世界，在这里，唯有演出才是焦点。礼堂共分为五层，拥有2,000个坐席，呈欧式马蹄铁形，是剧院的中心。它的设计主要考虑了三个因素：音响效果、私密性和优雅感。在歌剧中，清晰的台词和自然的声调十分重要，交响乐团的热情和共鸣也是必不可少的。为了保证最佳的音响效果，必须移除所有的背景噪音，而剧院正好位于喧嚣的地铁和有轨电车线路旁。因此，礼堂、舞台和排练厅都采用了独立结构。这个内部空间外围包裹着双层混凝土，坐落在500个高性能的橡胶垫上，完全符合一级声学标准，是加拿大第一座结构分离的表演设施。

1. Queen Street façade
2. Night view of the University Avenue façade
3. South-west façade detail

1. 皇后大街一侧外观
2. 学院大街一侧的建筑夜景
3. 西南侧外观细部

1

Below left: Orchestra (ground level)
Below right: Grand Ring (second level)

左下图：演奏大厅（一层）
右下图：大环形观众坐席（二层）

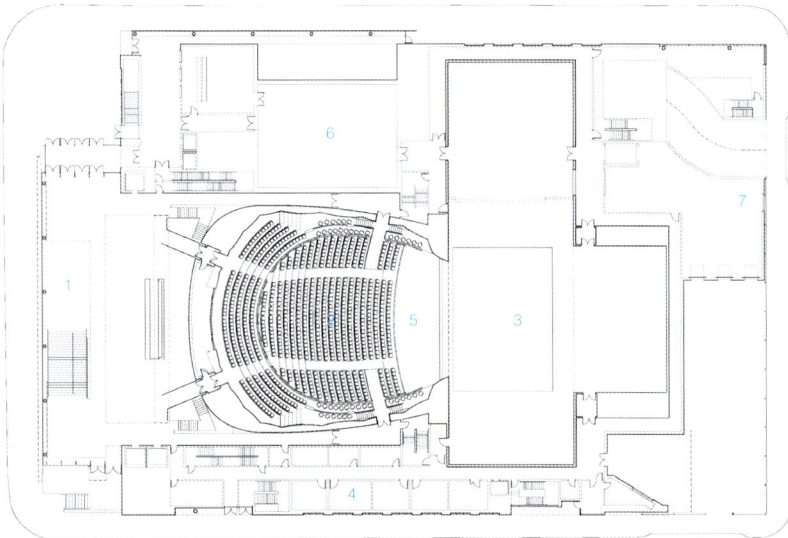

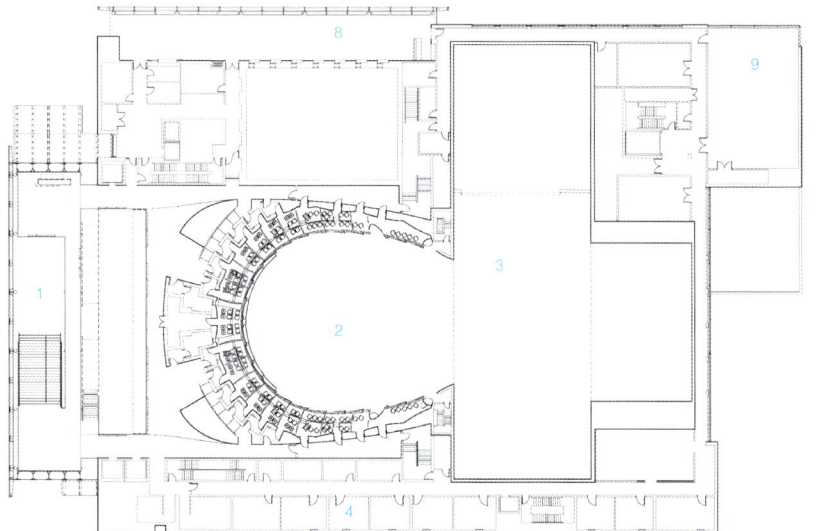

1. View from University Avenue median
2. Façade detail - a play of staircases in the City Room

1. 从学院大街中部看到的建筑外观
2. 外墙细部——城市空间中上演的一幕楼梯戏剧

Below: Third Ring (third level)
下图：第三层环形观众坐席

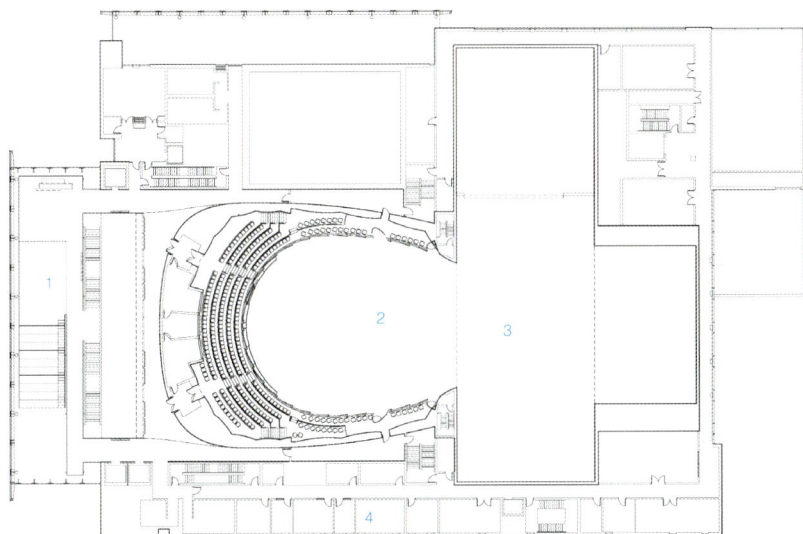

1. City Room
2. Auditorium
3. Stage/fly tower
4. Dressing room
5. Orchestra pit
6. Rehearsal room
7. Loading dock
8. Jackman lounge
9. Dance studio

1. 城市空间
2. 观众席
3. 舞台/舞台塔
4. 更衣室
5. 乐池
6. 排练室
7. 装卸处
8. 杰克曼休息室
9. 舞蹈工作室

1. Overview of the City Room
2. The glass stair with a horizontal run of almost 27 metres and a rise of 8.8 metres represents a significant development in structural glass design.
3. R. Fraser Elliot Hall

1. 城市空间全景
2. 玻璃楼横长27米，高度为8.8米，体现了玻璃结构设计的重大进步
3. R·弗雷泽·埃利奥特厅

剧院和音乐学校 # Theatre and Conservatory of Music

Location: Châtenay-Malabry, France **Designer:** Agence Nicolas Michelin & Associés (ANMA)
Completion date: 2008 **Photos©:** Stéphane Chalmeau **Area:** 8,077 square metres

项目地点：法国，沙特奈马拉布里 设计师：尼古拉斯·米其林建筑事务所 完成时间：2008年 摄影师：斯蒂芬那·沙尔莫 面积：8,077平方米

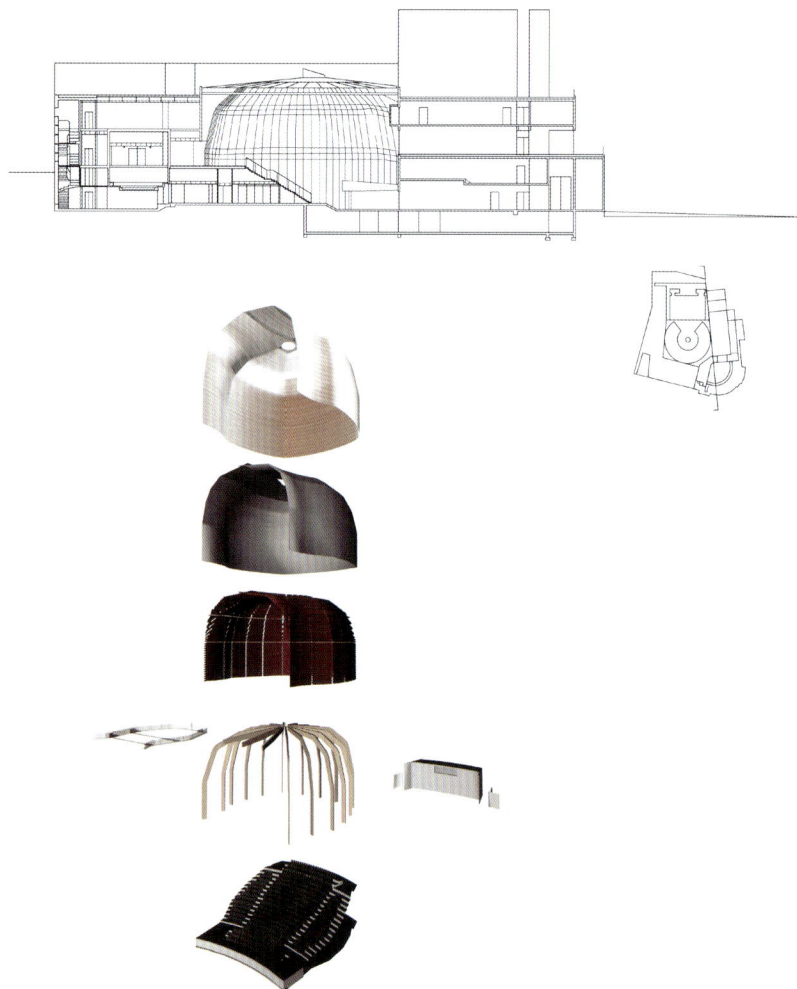

The programme of this cultural complex groups together three distinct entities: the theatre, the music and dance conservatory, and the music café. The project is developed around a centred plan to create a unitary building that communicates with the existing architecture – the old wing of the incineration plant that has been preserved to house the foyer. This composition expresses the diversity of the cultural complex's activities while also marking its unity.

The theatre stands at the heart of a complex formed by the former factory and the new "cascading" volumes around it that house the theatre lobbies, the conservatory rooms and the contemporary music rooms. The circulations are linked by an indoor street with a coloured glass roof that runs between the auditorium and the buildings of each entity.

The theatre seats a maximum of 550 people and forms a landmark for users of the cultural complex. Its walls are cladded in wood panels that integrate the acoustic treatment. The technical areas – for storing scenery and equipment are located around the stage house, with a direct link to the delivery bay. The rehearsal room is set behind the stage house, with the catering area and dressing rooms above it. The conservatory and its auditorium are housed in the extension wing, which closes off the courtyard. The façades are treated with a coloured plaster that echoes the neighbouring Butte Rouge garden city. The golden volume of the stage house indicates the stage equipment. Outside, the different levels are unified by use of the same material, thus creating a special perspective in the linear development of the new urban boulevard.

项目集合了三个独立的设施：剧院、音乐舞蹈学院和音乐咖啡厅。建筑的设计要求新建筑与原有建筑形成一个统一的整体，让旧焚化厂成为项目的大厅。这一结合体现了文化活动的多样性，也标志了建筑的统一性。
剧院位于建筑的中心，由工厂和环绕它的新空间组成。新空间包括剧院大厅、音乐学校教室和现代音乐室。拥有彩色玻璃屋顶的内部街道连接了礼堂和楼内的其他设施。
剧院最多能容纳550名观众，是一座建筑中的标志性元素。剧院的墙壁上镶着木板，以实现音响效果。用于储藏背景道具和设备的技术区环绕着后台，与配送区直接相连。排练室位于后台的后面，上方是餐饮区和更衣室。音乐学校和礼堂位于扩建的翼楼里，与庭院隔绝。外墙上涂抹着彩色石膏与周围的建筑遥相呼应。后台的黄金空间里摆放着舞台设施。外部的各个楼层都采用了同种材料，在城市环境中形成了一种特别的景观。

1. Existing building façade　　1. 原有建筑外观
2. Main façade　　　　　　　2. 新建筑外观
3. Entrance　　　　　　　　3. 入口
4. Exterior general view　　　4. 建筑全景

1

2

3

4

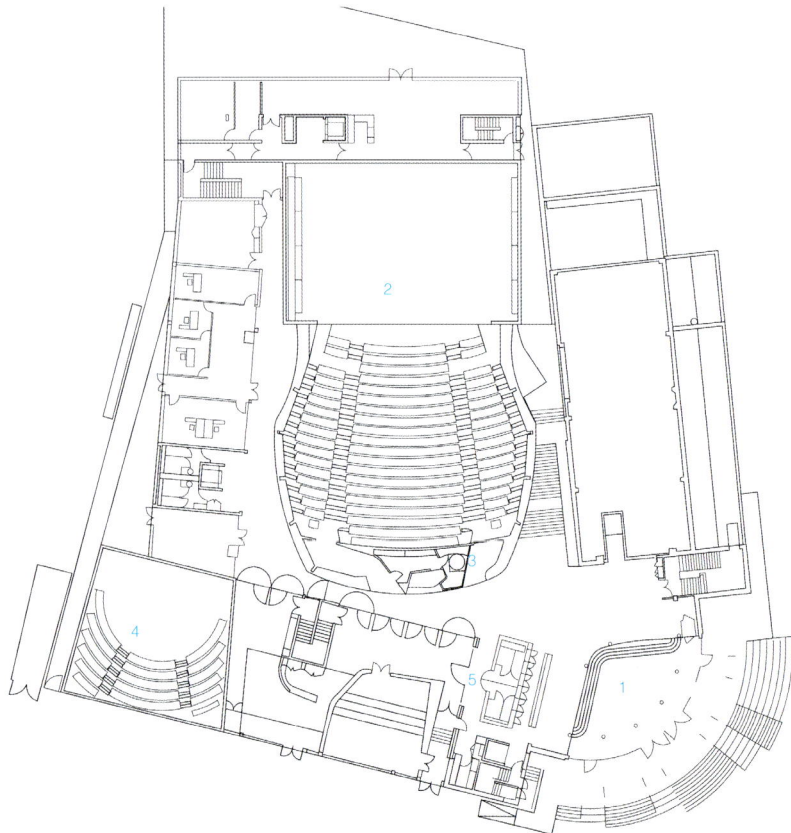

1. Entrance
2. Stage
3. Main hall
4. Music hall
5. Toilet

1. 入口
2. 舞台
3. 主厅
4. 音乐厅
5. 洗手间

1. Roof detail
2. Interior detail
3. Entrance of the theatre

1. 屋顶细部
2. 室内细部
3. 剧院入口

1

2

1. The old wing of the incineration plant that has been preserved to house the foyer
2. Rehearsal room
3. Stairway

1. 焚化厂的旧翼楼被改造为门厅
2. 排练室
3. 楼梯

3

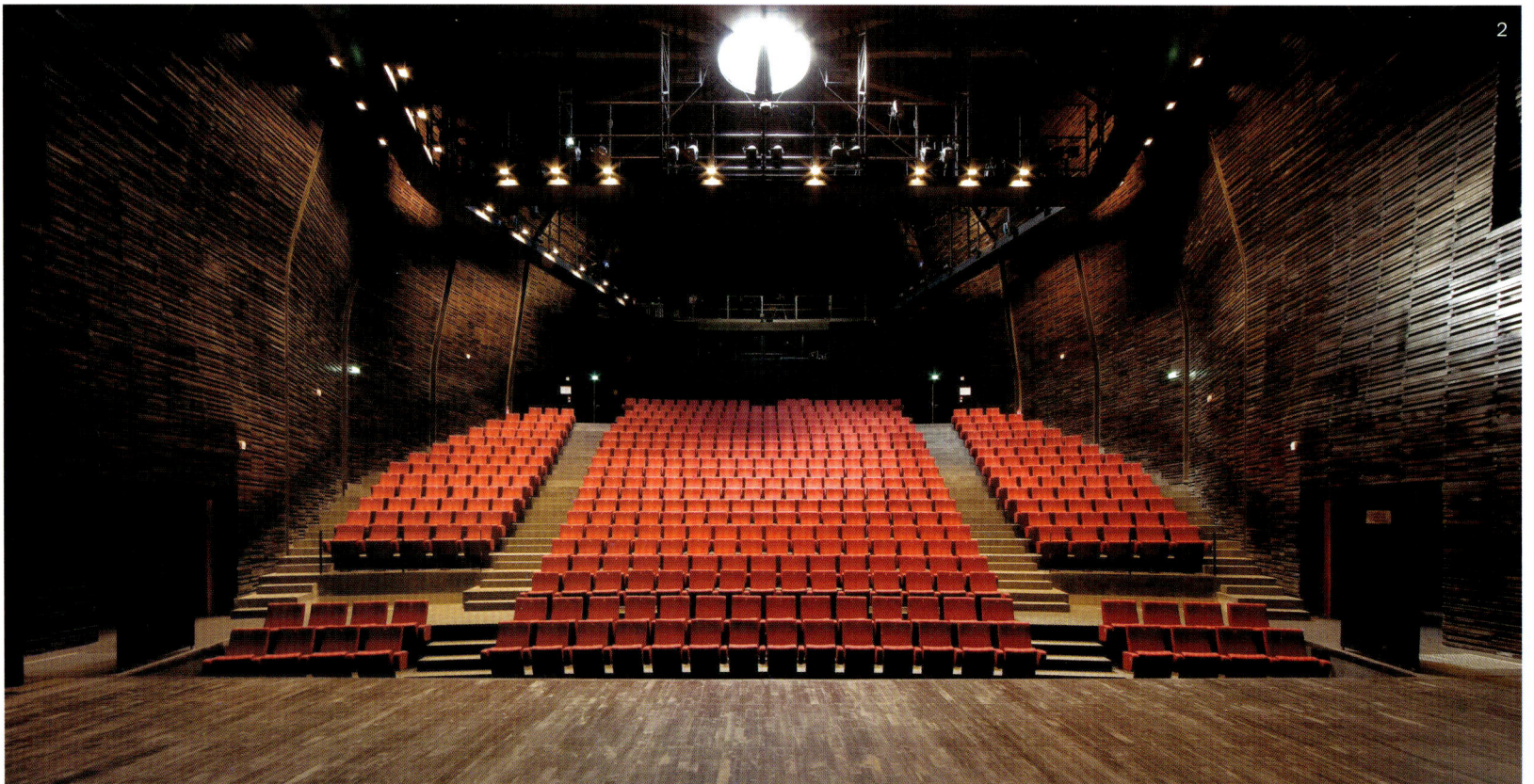

1. Stage view
2. Auditorium
3. Music hall

1. 舞台
2. 剧院观众席
3. 音乐厅

柯达伊中心 # Kodaly Centre

Location: Pécs, Hungary **Designer:** Építész Stúdió **Completion date:** 2010 **Photos©:** Tamás Bujnovszky
Area: 11,200 square metres

项目地点：匈牙利，佩奇 设计师：艾皮特兹工作室 完成时间：2010年 摄影师：托马斯·布吉诺维斯基 面积：11,200平方米

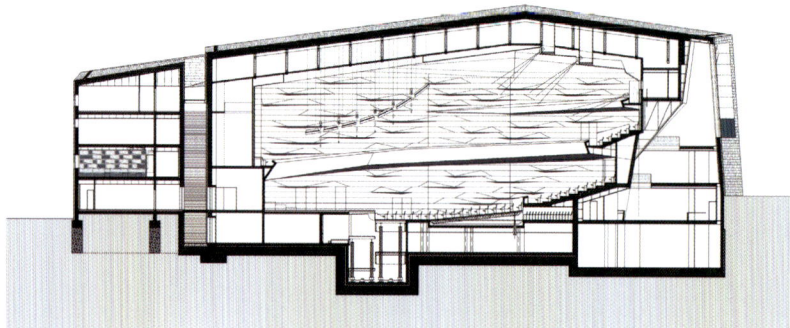

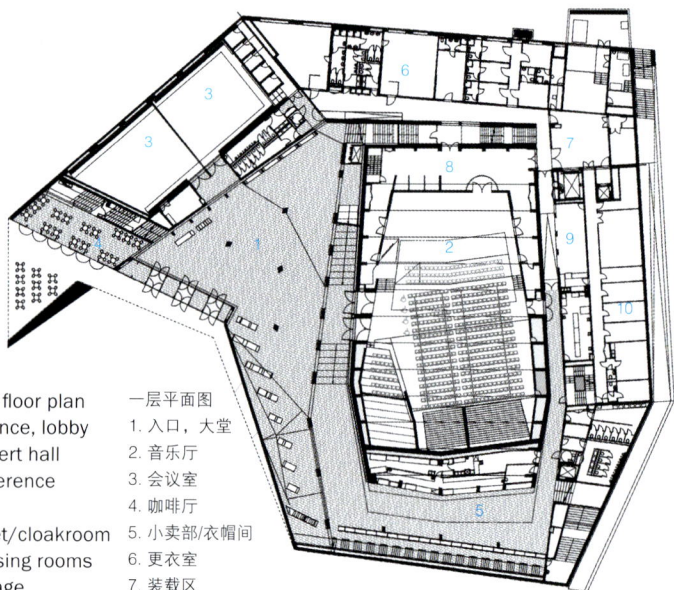

The Hungarian city of Pécs was selected as European Capital of Culture for 2010. The new Concert and Conference Centre is one of the main projects for this event.

The network of musical institutions in Pécs appears to be quite complete as far as training and performers are concerned. However, there was no concert hall that could host performances in a worthy manner. For this reason the goal of this project is to establish an internationally significant, acoustically designed multifunctional building with modern background technology that operates as a concert hall and a conference centre.

Once it is built, the internationally renowned symphonic orchestra of the region, the Pannon Philharmonic and several other musical ensembles of the city will be able to continue their successful work, and thanks to the cultural space to be also established, Pécs will be able to offer a much wider range of cultural opportunities. The conference function will make Pécs a significant middle-sized venue of the conference market in Central Europe and the city will be able to host professional conferences, fairs and cultural festivals. With the related investments (motor way, regional airport) the competitiveness of Pécs will improve significantly in cultural and conference tourism.

The new building include, in addition to a concert hall and a large rehearsal room, the offices of the Pannon Philharmonic and the Conference Centre, other rooms necessary for the operation of the orchestra (such as storerooms for sheet music and instruments), facilities serving the audience – café, bookstore, lounge, etc. – and several service premises.

佩奇被选为2010年欧洲文化之都，新建的音乐厅和会议中心是其中的主要项目之一。

就培训和表演者而言，佩奇的音乐机构网络十分复杂。但是，城市内并没有一所知名的音乐厅。因此，项目的目标是结合现代背景技术，打造一座兼具音乐厅和会议中心的国际知名的多功能建筑。

音乐厅一建成，闻名国际的当地交响乐团——潘诺交响乐团和一些其他音乐团体将可以在此演绎作品。而且，佩奇将能够提供更多的文化机遇。会议功能将使佩奇在中欧的会议市场中占有一席之地。佩奇将能够举办专业会议、展会和文化节。配合相关设施（如高速公路、地方机场等），佩奇在文化和会议旅游方面的竞争力将得到巨大的提高。

新建筑除了包含一个音乐厅和一个巨大的排练室之外，还有潘诺交响乐团的办公室和会议中心、交响乐团所必需的其他运营空间（如乐谱和乐器仓库）、观众服务设施——咖啡厅、书店和休息室等和一些基础服务设施。

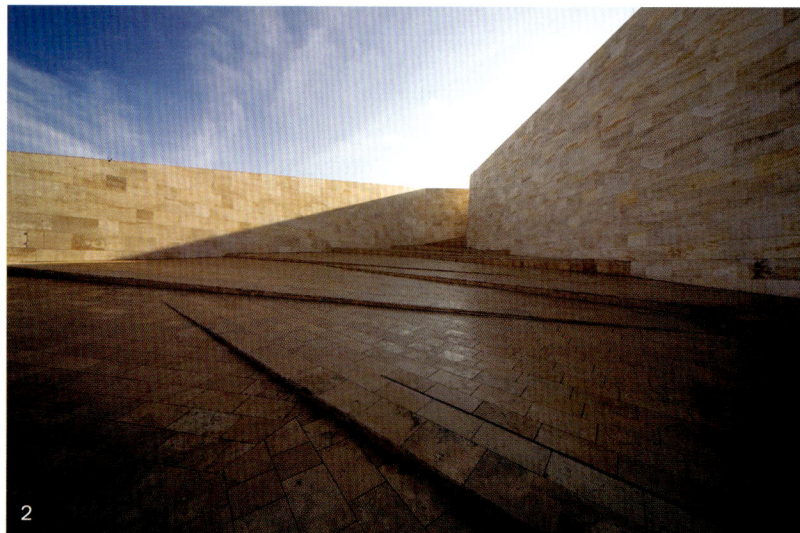

Ground floor plan / 一层平面图
1. Entrance, lobby — 1. 入口，大堂
2. Concert hall — 2. 音乐厅
3. Conference — 3. 会议室
4. Café — 4. 咖啡厅
5. Buffet/cloakroom — 5. 小卖部/衣帽间
6. Dressing rooms — 6. 更衣室
7. Stowage — 7. 装载区
8. Backstage — 8. 后台
9. Canteen — 9. 食堂
10. Service — 10. 服务区

1, 2. Façade detail — 1、2. 外观细部
3. Side exterior façade — 3. 建筑侧面

1

2

1. Entrance plaza
2. General façade
3. Façade night view

1. 入口广场
2. 建筑全景
3. 建筑外观夜景

First floor plan
1. Foyer
2. Concert hall
3. Conference/ballet room
4. Rehearsal rooms/section rooms
5. Artist entrance
6. Administration

二层平面图
1. 门厅
2. 音乐厅
3. 会议室/芭蕾室
4. 排练室/部门办公室
5. 演员入口
6. 行政区

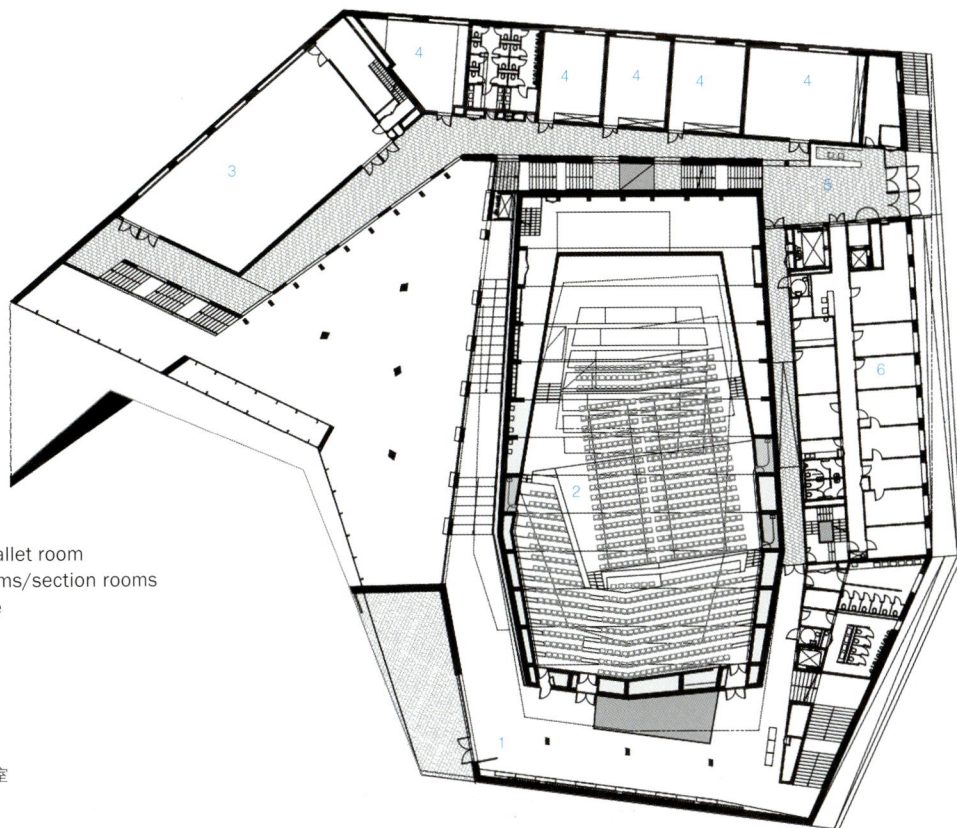

1

1. Lobby/reception
2. Hallway

1. 大堂/前台
2. 走廊

2

1. Concert hall
2. View showing stage from the upper seating
3. Concert hall detail

1. 音乐厅
2. 从上层坐席看舞台
3. 音乐厅细部

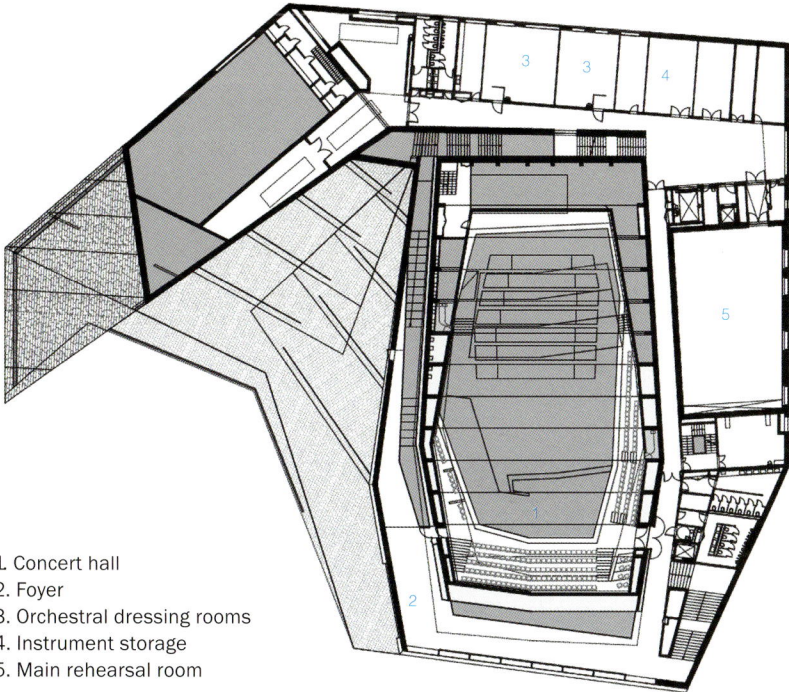

1. Concert hall
2. Foyer
3. Orchestral dressing rooms
4. Instrument storage
5. Main rehearsal room

1. 音乐厅
2. 门厅
3. 乐队更衣室
4. 乐器储藏室
5. 主排练室

中国电影博物馆 **China National Film Museum**

Location: Nanjing, China **Designer:** RTKL **Completion date:** 2007 **Photos©:** Yuan, Hsiaonan

项目地点：中国，南京 设计师：RTKL 完成时间：2007年 摄影师：袁晓楠

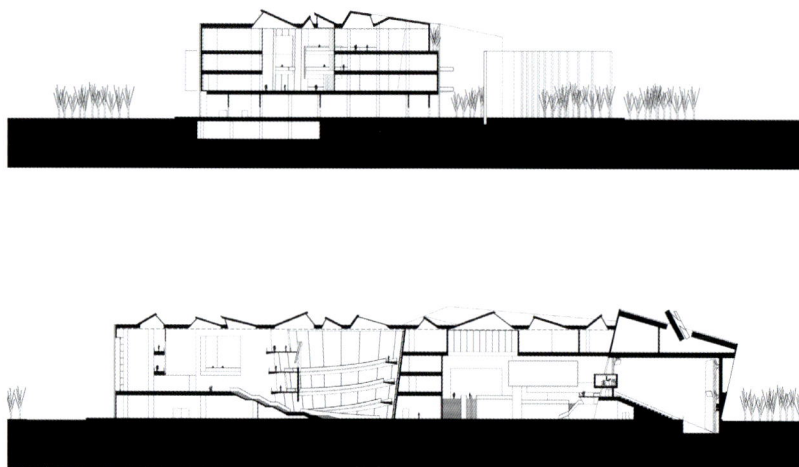

Like the nation itself, China's film industry has evolved dramatically over the last century. The museum's programmes will introduce China's rich cinematic traditions to a new generation of international audiences. The goal of the design was to create a not only iconic but also experiential architecture that is dedicated to, generated from and interactive with the special object of film, and also connotes an institution of a particular cultural context.

At 38,500 square metres, the building comprise four levels of exhibition halls for film history, film technology and temporary exhibits; a cinema complex composed of various theatres, a multifunction hall; shop and restaurant; collection and storage; and research and administrative offices, all within a single volume standing on an open site in the city's outskirt.

Associated with the architecture–film duality is a group of interrelated questions such as what's institutional versus what's entertainment; what's permanent versus what's ephemeral; elite art versus popular culture; virtual sensation versus real experience, etc. These dynamic parallels are reflected in the synthetic and layered design approach that aims to catalyze an unconventional and enlightening experience by bringing together distinctive characteristics of film and architecture.

The multi-fold design task is translated into a building that is both grand and versatile in imagery and spaces.

Monumentally scaled to match the open context, the building reserves all its programmes and the excitements within a singular black rectangular box. While the building appears monolithic from a distance, it gradually reveals its translucency and openness upon approaching through its perforated skin treatment and entry statement. An arrival experience is sequenced with a series of outdoor and indoor spaces to allow a layered expression and stage zoning to manage crowds going in and through the building.

An interior street penetrates the buildings and enables free public access to a significant portion of the building's interior space and venues, creating an urban atmosphere as a prelude to the museum's internal experience. Contrast greatly with the singularity of the building envelope, the interior is a collection of various interconnected spaces that are staged to deliver a montage experience, which is also reminiscent of roaming in Chinese gardens known for their fixed vistas and spaces for improvisational viewings.

To claim the nature of "popular art" in film and a grassroots aesthetic, the design resorts to recognisable visual icons of cinemas, ranging from the universal cinema icon, the production clapboard, as the overall architectural form and statement, to the giant star-shaped entrance, and graphical application of millions of mini star perforations populating the entire building envelope.

In addition to tangible forms, lighting plays a critical role in rendering atmosphere and space-making. Natural lights into the black box are filtered or stylishly tinted with primary colours. Theatrical fixtures are used in lieu of conventional lighting. Screens are transformed as architectural and lighting tool in crafting an environment, as in the central rotunda that is both theatrical and architectural.

When architecture encounters and merges with another entity, in this film, it evolves.

1. Main façade night view　　1. 建筑外观夜景
2. Façade detail　　　　　　 2. 外观细部
3. Art installation on the plaza　3. 广场上的艺术装置

1

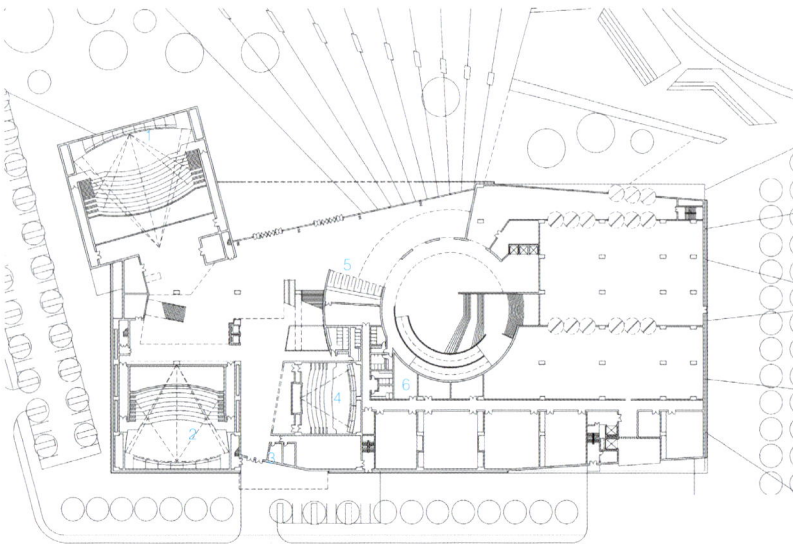

近一个世纪以来，中国电影业同中华民族一样经历了惊天动地的变化。博物馆将向新一代的各国电影观众展示中国丰富的电影传统。设计的目标不仅仅是要构思出一座标志性的建筑，更要构思出一座体验性的建筑，源自电影、服务于电影、与电影相互促进，同时这里也是这一特殊文化的研究机构。

项目位于城市郊区一片开阔地带，占地3,577平方米，包括四层展厅，分别用于电影历史、电影科技展览和临时展出；还包括有多个放映厅的电影院、多功能厅、店铺和餐厅、藏品收藏室，以及研究和管理室，都安排在这座独栋建筑内。

同建筑和电影的二元性相关的，是一系列关联性问题，比如机构性场所与娱乐性场所、永久性与短暂性、高雅艺术与通俗文化、虚拟感受与真实体验等等。这些动态平行的元素通过一种综合而分层的设计手法得以体现，将建筑和电影的鲜明特征加以整合，以求催生出一种异于传统的启蒙意味的体验。

恢宏而功能完备的建筑形态和空间形式，充分体现出设计的多层次性。

建筑物以其巨大的形体配合空旷的背景，呈现为一个单体的黑色方盒，容纳所有的建筑内容和功能。从远处看，它高高耸立；逐渐靠近后，通过其镂空的建筑外壳和入口设计，渐渐显现其半透明的形态和开放空间。自到达点起，室内外形成一系列空间，确保设计理念的层次性；分区设计对进出建筑的人流进行引导。

贯通室内的街道为公众提供连通内部各重要空间和场所的便捷的交通，使得参观者一踏足博物馆，即感受到城市氛围。与建筑外观的特异性明显不同的是，内部由一系列相互连接的空间构成，阶段性地给人以蒙太奇般的体验，同时让人忆起漫步于以移步换景风格著称的中国园林中的感觉。

为表现电影中"通俗艺术"和大众美学的趣味，设计上借用了通俗易懂的电影图形符号，包括通用的电影符号和拍摄对号板作为整体的建筑形式和建筑语言；采用了巨大的星形大门，并在整个建筑外表上采用成千上万的小星星穿孔图案，形成整体建筑外观；

除了这些实体的形式外，照明在营造气氛和布设空间上也起着很大的作用。照进黑盒子的自然光进行过滤或调色处理成了原色。用舞台布景取代了传统的照明。帷幕也被调用，作为建筑和照明工具，参与场景的布置，就比如在中央圆形大厅那样，剧场和建筑的功能得到了和谐的配合。

当建筑与另外一个实体（在这里是电影）相遇并融合时，它便会慢慢演进。

1. Giant-screen cinema   1. 巨幕电影院
2. 4D stereoscopic cinema   2. 4D立体电影院
3. Guard/restroom   3. 门卫/卫生间
4. Small hall   4. 小型观摩厅
5. Ticket office   5. 售票区
6. Administration office   6. 行政办公

1. Façade detail
2. Lobby

1. 外观细部
2. 大堂

2

1

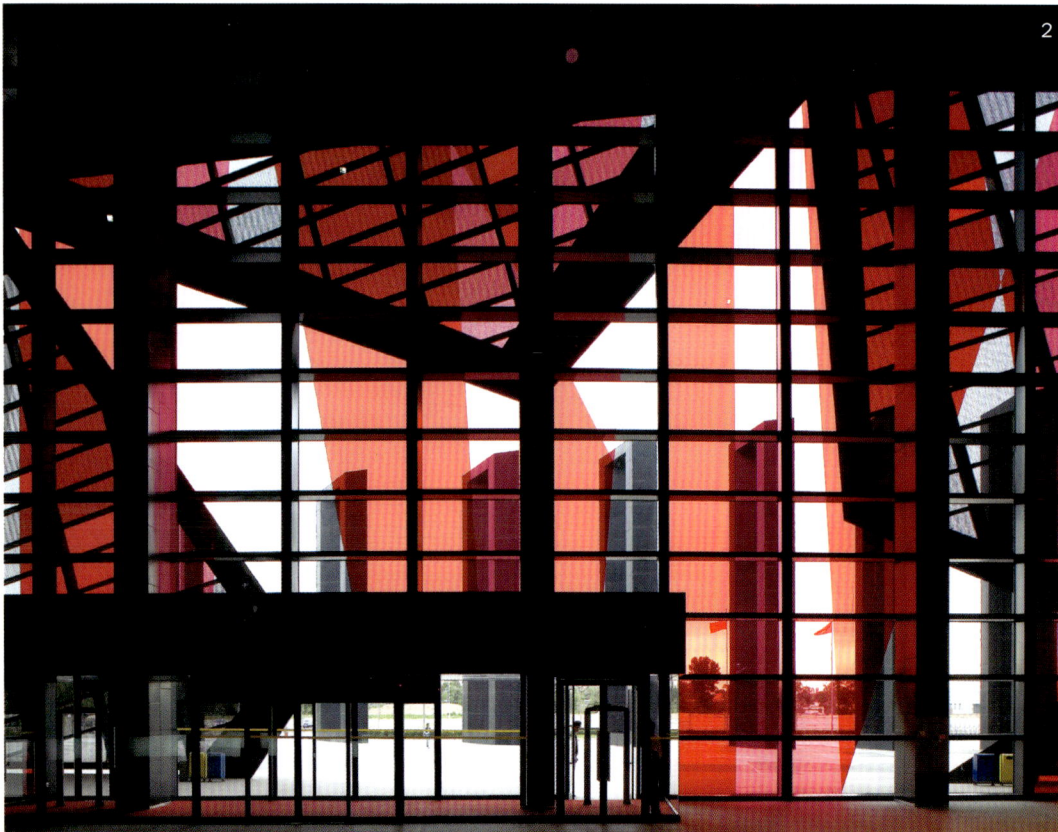

1. Lobby/information service
2. Entrance and façade detail viewed from the lobby
3. Top view of the lobby

1. 大堂/信息服务中心
2. 从大堂看入口和外观细部
3. 大堂顶部

3

1. IMAX giant-screen cinema
2. Top of multi-function hall
3. Science and educational film hall
4. News and documentary hall
5. Sub room
6. Top of the medium hall
7. Film poster hall
8. Cartoon hall
9. Hong Kong and Macau film hall
10. Dubbed film hall
11. Taiwan film hall

1. IMAX巨幕电影院
2. 多功能厅上空
3. 科教片厅
4. 新闻纪录片厅
5. 辅助用房
6. 中厅上空
7. 电影海报厅
8. 美术片厅
9. 港澳厅
10. 译制片厅
11. 台湾厅

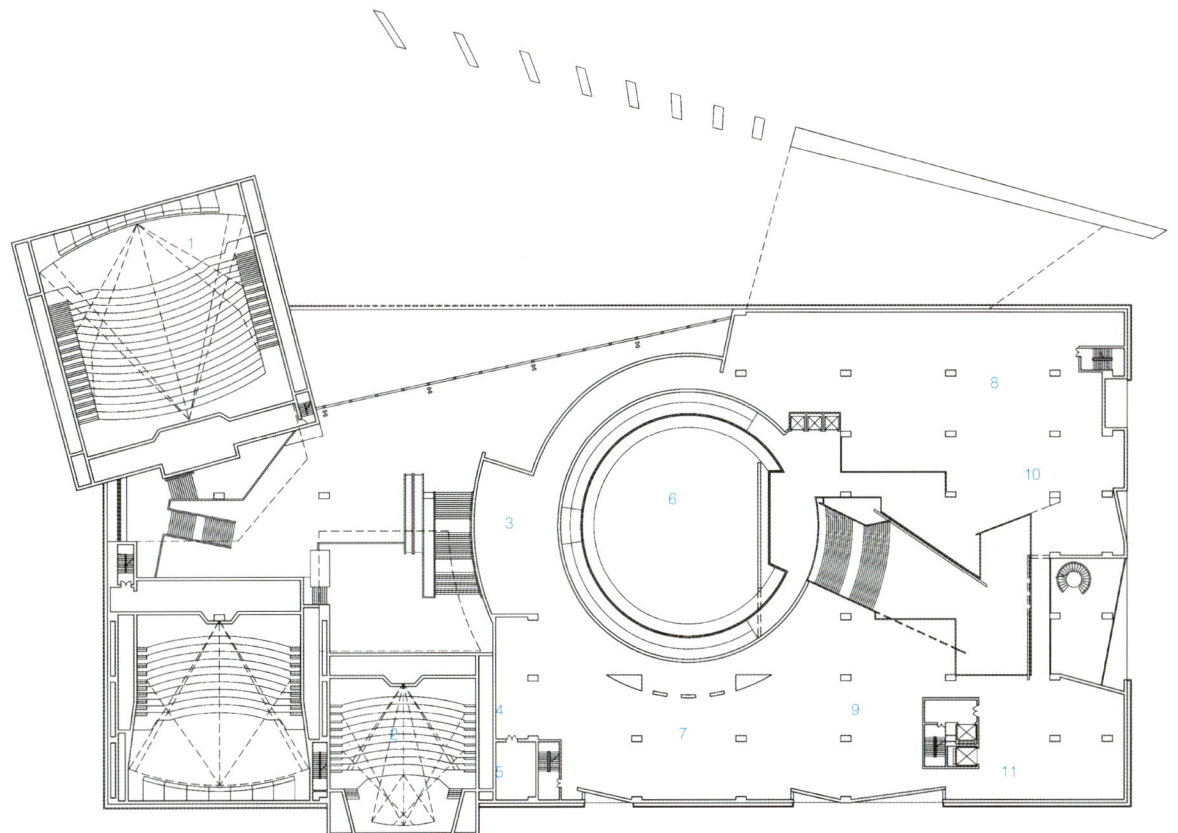

1, 2. Cinema hall
3. Small hall

1、2. 影厅
3. 小观影厅

穆尔斯卡索博塔剧院公园 Theatre Park In Murska Sobota

Location: Murska Sobota, Slovenia **Designer:** Andrej Kalamar, Studio Kalamar **Completion date:** 2010
Photos©: Miran Kambic **Area:** 1,460 square metres

项目地点：斯洛文尼亚，穆尔斯卡索博塔 设计师：安德烈·卡拉马尔、卡拉马尔工作室 完成时间：2010年 摄影师：
米兰·凯姆比克 面积：1,460平方米

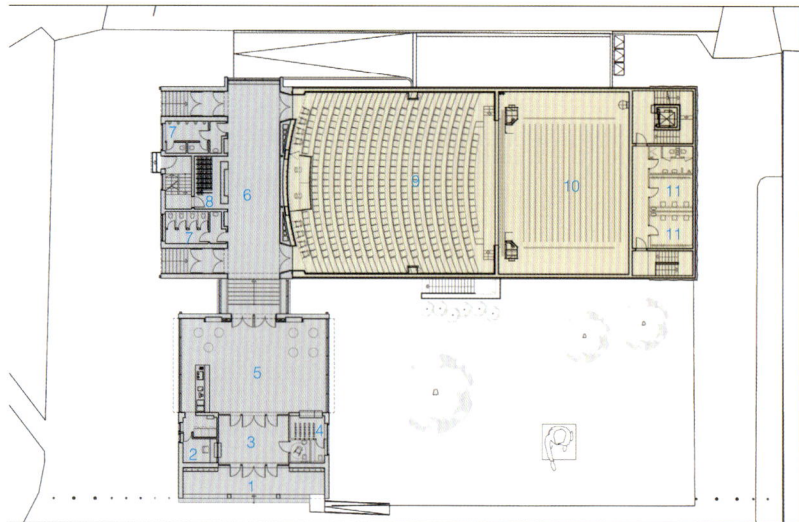

The Cinema Park Building was erected in the 1950s from plans by the modernist architect Franc Novak. Slowly falling into disrepair, it was still used for film shows until the new century, when it became obsolete due to the construction of a new multiplex. A solution was found in a new, more ambitious programme: a conversion into a theatre and concert hall, preserving the designated landmark modernist building with a fitting new programme.

In relation to the existing structure, two concepts were introduced for the two segments of the project. The functional scheme of entry areas already included all the key elements: entrance loggia, lobby and the lower and upper foyers. These parts of the building were renovated, preserving most of the original substance. The existing auditorium's volume was not suited to the new purpose, so it was replaced by a new, correctly proportioned auditorium space and a technically equipped theatre stage large enough to house performances from every theatre in the country. The volume of the original building is clearly delineated in the auditorium's structure, the added stage volume shows an appropriate compositional distance and retains the autonomy of the original building's architectural expression.

影院公园建于20世纪50年代，由现代建筑师弗兰克·诺瓦克设计。影院一直用于放映影片，直到21世纪，一座新的影院建成。新规划计划将影院改造成一座剧院和音乐厅，让这座标志性的现代主义建筑重获新生。
项目在原有结构的基础上，引入了两个全新的概念。入口区域的功能设计包含了所有的关键元素：门廊、大厅和高层、底层门厅。这些部分都经过了整修，保留了原汁原味。由于原来礼堂的规模不适合新功能，因此，设计师用一个全新的、比例适当的礼堂和设施齐全的剧院舞台替代了它。新礼堂能够上演国内的任意剧目。原有建筑的体量在礼堂结构中得以清晰地体现，添加的舞台空间距原有建筑有适当的距离，保证了原有建筑的独立性。

1. Entrance loggia
2. Tickets
3. Lobby
4. Cloakroom
5. Lower foyer
6. Upper foyer
7. Restrooms
8. Cloakroom
9. Auditorium
10. Stage
11. Dressing rooms of actors

1. 入口凉廊
2. 售票处
3. 大堂
4. 衣帽间
5. 休息大厅
6. 上层休息大厅
7. 休息室
8. 衣帽间
9. 剧院礼堂
10. 舞台
11. 演员更衣、化妆室

1. Western elevation, seen from street
2. View of back elevation from the park
3. Front elevation
4. View from the street, entrance loggia and foyers at the front

1. 从街道一侧看建筑的西立面
2. 从公园一侧看建筑的后立面
3. 正面
4. 从街道一侧看到的入口凉廊以及正面大厅

1

2

1 View from the street
2. West and south elevations, viewed from the park
3. Detail: stage volume, with Second World War Memorial at the front

1. 从街道一侧看建筑
2. 从公园看西侧和南侧立面
3. 细部：舞台一侧的建筑体量，前方配有第二次世界大战纪念碑

1. Stage volume – facade detail: metal "ribs"
2. Entrance, ticket office
3. Lower foyer – view to the entrance

1. 舞台一侧的建筑体外立面细部：金属骨架
2. 入口、售票处
3. 从一层大厅看入口

1

2

3

4

1. Lower foyer – view of the garden side
2. Lowe foyer –bar with view of the park behind
3. Upper foyer, view towards the park
4. Upper foyer, seating detail

1. 一层大厅伴随花园景色
2. 一层大厅、吧台及其后的公园
3. 从高处的大厅看向公园
4. 高处大厅的座椅细部

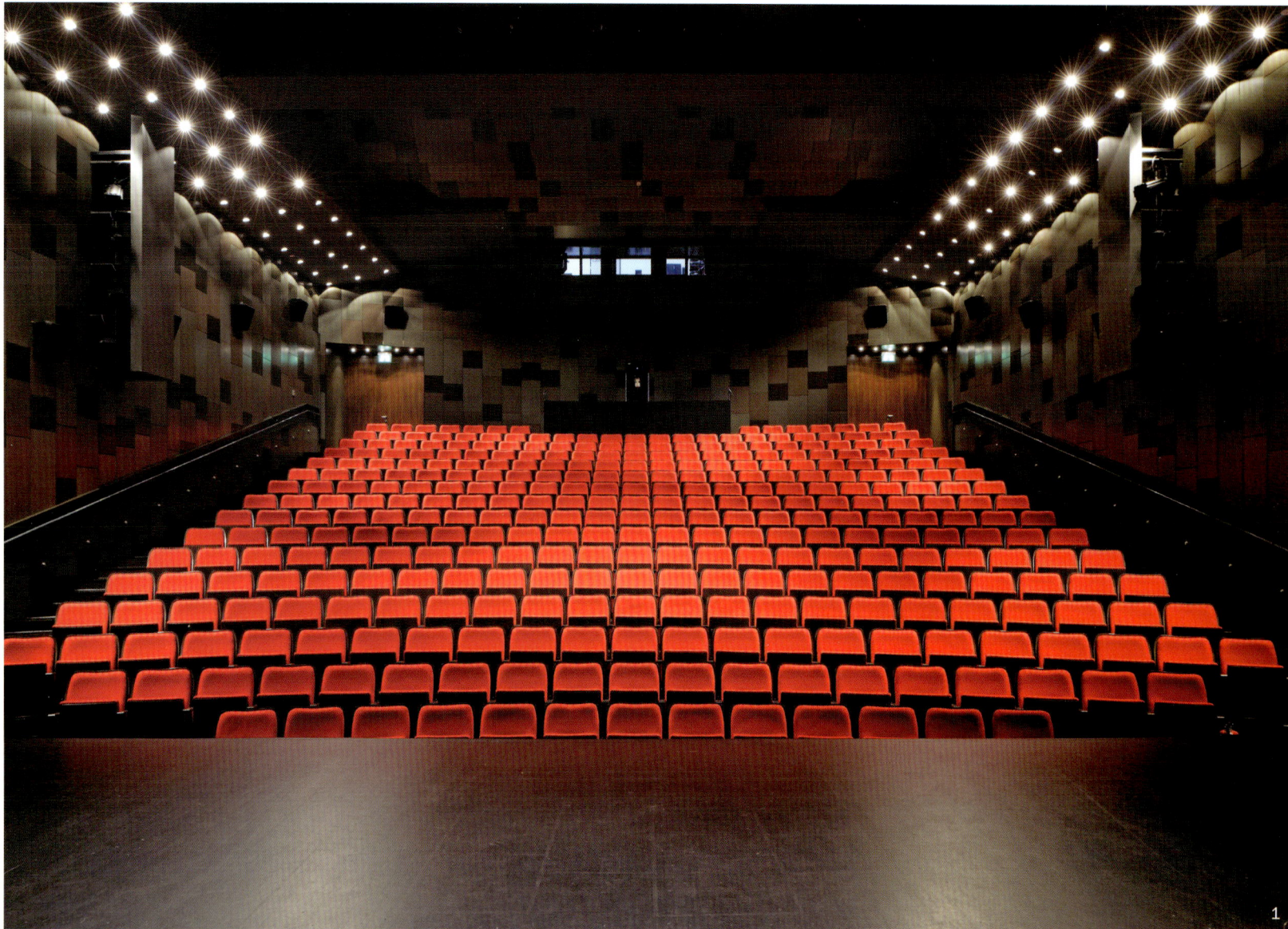

1. Auditorium, view from stage
2. Auditorium, view towards the stage
3. Auditorium, acoustic panelling detail

1. 从舞台看观众席
2. 从观众席看向舞台
3. 剧场隔音板细部

音乐剧院 # Music Theatre

Location: Graz, Austria **Designer:** UNStudio **Completion date:** 2008 **Photos©:** Iwan Baan, Christian Richters **Construction area:** 6,200 square metres **Award:** ULI 2010 Award for Excellence/2010 IIDA Annual Interior Design Award

项目地点：奥地利·格拉茨 设计师：UN工作室 完成时间：2008年 摄影师：伊万·巴恩、克里斯汀·里奇特尔斯 建筑面积：6,200平方米 所获奖项：2010年美国城市土地利用学会优秀奖；2010年国际室内设计协会室内设计奖

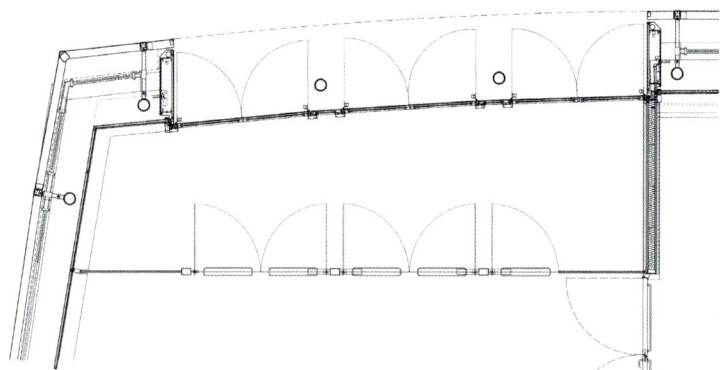

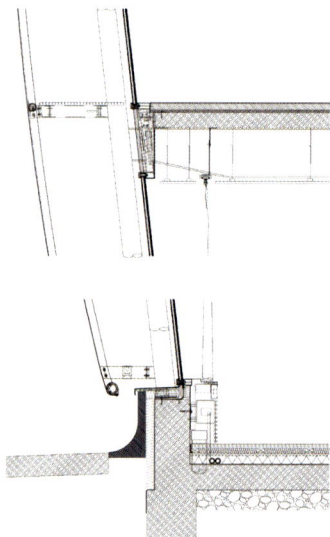

In the Music Theatre in Graz, the unit-based part of the organisation (the box) is situated on the right side and the movement-based part (the blob) on the left side of the building as seen from the Lichtenfelsgasse. There are two entrances: the everyday entrance on the park side which is used by students and staff, and the public entrance on the Lichtenfelsgasse which is used by the audience when there is a performance. On performance nights, the student entrance is transformed into a wardrobe using mobile closets. A removable ticketing desk and screen bulletin are placed underneath the staircase. The public ascends a wide staircase and enters a large foyer on the first floor. This foyer gives access to the multipurpose auditorium that can seat up to 450, and that is adaptable to a great variety of performances, ranging from solo instruments to opera and to full orchestra.

The free-flowing space of the foyer is made possible by a spiraling constructive element that connects the entrance to the auditorium and to the music rooms above, thus welding together "with a twist" the three levels of this side of the building. The twist is in fact a massive concrete construction, which was one of the most challenging UNStudio ever realised – more difficult to achieve even than the twists in the museum for Mercedes-Benz. The dimensions of this particular twist necessitated far greater precision and the use of self-compacting concrete, which was pumped up from below instead of poured down from above as is the usual method. The twist forms a central feature of the public space, around which everything revolves. Lighting and material details accentuate the ripple effect. The twist is highlighted from above by a skylight in the ceiling, which itself consists of lamellas executed in dark wood that fan out from the twist in a wave-like pattern.

With the overt presence of the spring receding from the façade as the design evolved, the exterior again became a blank canvas, generating the opportunity to return to the theme of music in a new way. The interest in re-establishing a relationship between music and architecture had from the beginning focused on shared aspects such as rhythm, continuity, and channeling. Through the readings of the philosopher Gilles Deleuze that there is another element that had not been seriously studied before: the element of repetition. Repetition generates an aggregate with densifications, intensifications and intervals. Repetition brings sonority. It allows for improvisation, it marks territory, it codes milieus. The designers decided to use a repetitive pattern, of the design, and apply this to the façades in various ways to achieve some of these effects. The pattern, executed in the muted tones of stage make-up, is found all over the building in various degrees of density. Its appearance is furthermore impacted by changes in light during night and day, as well as by proximity and view angles since the outermost layer of the façade consists of a glittering mesh.

从里奇特福尔斯加斯看去，以单元为基础的盒结构位于建筑的右侧，而以移动为基础的团状结构位于建筑的左侧。建筑有两个入口：日常入口朝向公园，供学生和工作人员使用；公共入口朝向里奇特福尔斯加斯，供观众观看演出时用。在演出之夜，学生入口将变成一个运用可移动衣柜的更衣室。移动式售票处和公告栏将设在楼梯下方。公众通过宽大的楼梯进入二楼的大厅。大厅通往可容纳450人的多功能礼堂，可以进行乐器独奏、歌剧、交响乐团等多种类型的表演。

自由流畅的大厅通过盘旋的结构元素连接着入口、礼堂和音乐室，从而将三层楼扭曲地联系在一起。这一扭曲的结构其实是巨大的混凝土结构，十分具有挑战性，甚至比梅赛德斯-奔驰博物馆的扭度还要难以实

1. Curtain-wall detail 　1. 幕墙细部
2. Building exterior 　　2. 建筑外观
3. Overall view 　　　　3. 建筑全景

现。这个特别的扭曲结构的尺寸需要更高的精确度，自我压缩的混凝土是从下至上由泵压抽上来，而不是自上而下浇铸而成的。扭曲结构是公共空间的中心元素，其他元素都围绕着它展开。灯光和材料的细部突出了涟漪效果。天花板上的天窗强调了扭曲结构。天窗由从扭曲结构中呈扇形散开的黑木薄层组成，形成了波浪图案。

随着外墙的存在感在设计过程中逐渐消退，建筑的外观再次变成了空白的帆布，以一种全新的方式回归了音乐主题。设计师想要在音乐和建筑之间重塑一种联系，二者在韵律、连续性、通道方面都拥有共同点。通过对哲学家吉尔·德勒兹的研究，设计师发现了前所未有的共同元素——重复。重复结合了致密化、集约化和间隔。重复带来宏亮。它考虑到即兴创作，它标志着音域，它编织着环境。设计师决定在建筑外墙设计上采用重复的图案，以达到以上效果。柔和的图案遍布建筑，时疏时密。由于建筑的最外层采用了闪闪发光的金属网，日夜照明条件、远近和视角的变化让图案更显紧凑。

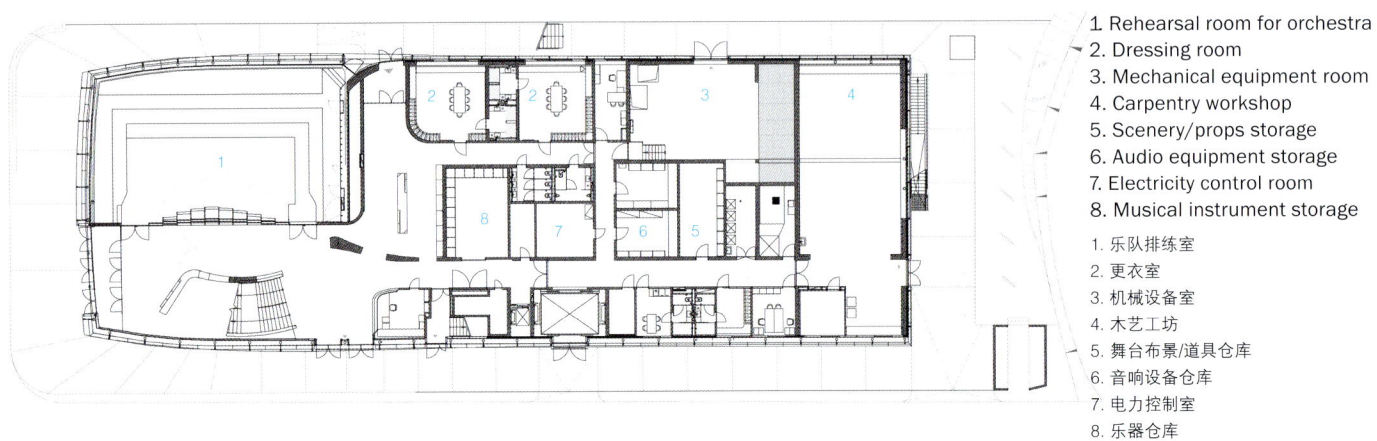

1. Rehearsal room for orchestra
2. Dressing room
3. Mechanical equipment room
4. Carpentry workshop
5. Scenery/props storage
6. Audio equipment storage
7. Electricity control room
8. Musical instrument storage

1. 乐队排练室
2. 更衣室
3. 机械设备室
4. 木艺工坊
5. 舞台布景/道具仓库
6. 音响设备仓库
7. 电力控制室
8. 乐器仓库

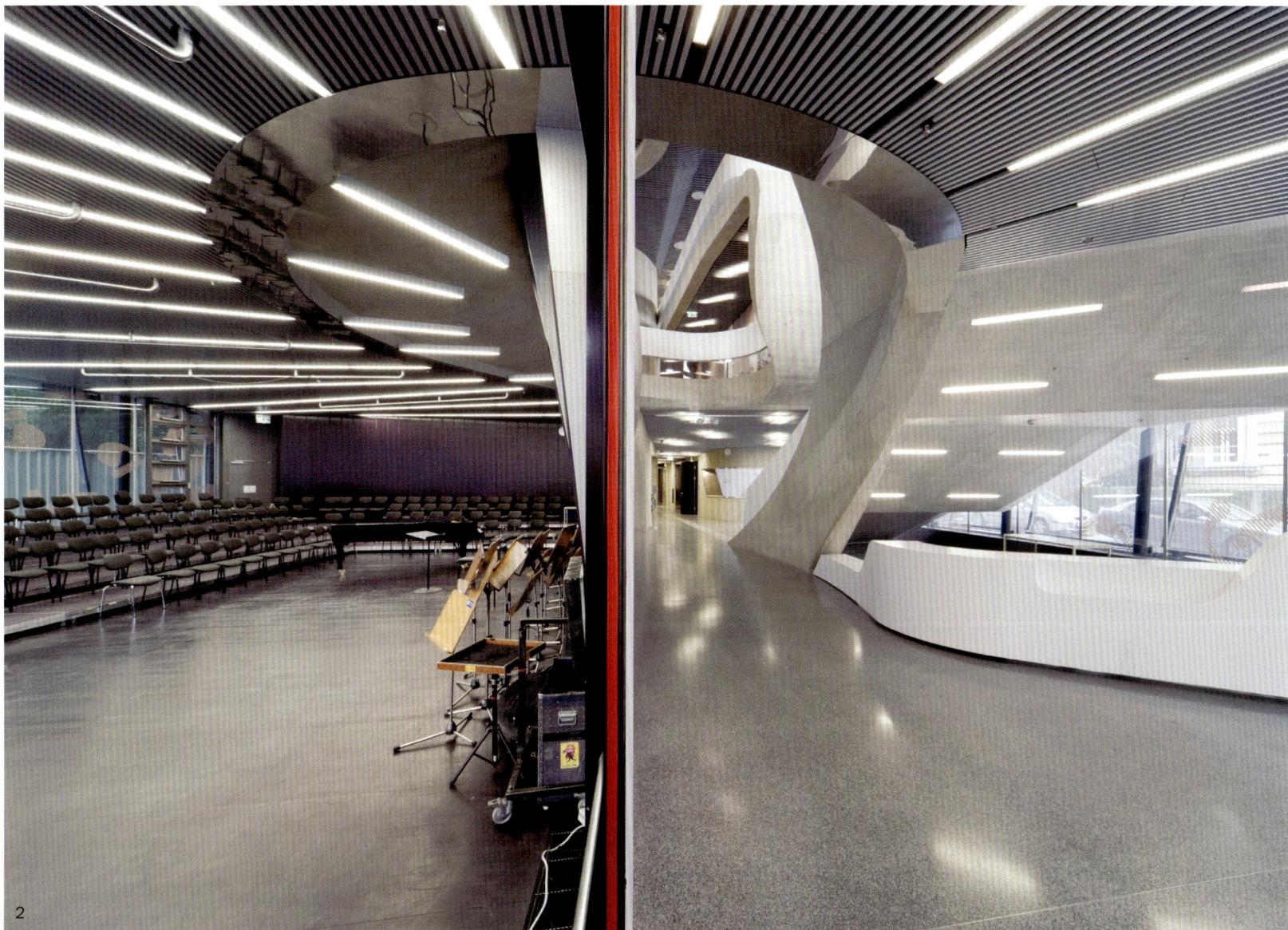

1. The structure of the twist as seen from the public foyer on the first floor
2. Ground floor dissection (left: orchestra rehearsal room; right: public entrance and lobby)
3. Lobby at public entrance on ground floor with custom-made reception desk, suspended ceiling lamellas and recessed lighting

1. 从二楼公共大厅看扭曲的结构
2. 一楼剖面（左：乐队排练室；右：公共入口和大堂）
3. 一楼的公共入口和大堂，定制的前台、悬挂式天花板和嵌入式灯

1. "Red Carpet" stairway and public entrance to black box theatre, in public foyer on first floor
2. Rehearsal room on the third floor with timber flooring
3. Restaurant/cafeteria on the second floor, with custom-made seating and service area

1. "红毯"楼梯和实验剧院的公共入口，位于二楼公共大厅
2. 四楼排练室，木地板
3. 三楼餐厅/自助餐厅，特别定制的座椅和服务区

1. The black box theatre contains custom-made acoustic panels to accommodate the varied music types performed in the faculty theatre
2. Dressing room on the third floor
3. Carpentry workshop on the ground floor
4. Fly gallery above black box theatre; walkways provide ease of access to stage machinery

1. 实验剧院拥有定制的隔音板以适应不同类型的音乐表演
2. 三楼更衣室
3. 一楼木艺工坊
4. 实验剧院上方的布景长廊；通道便于舞台装置的进出

马克·泰帕论坛 # Mark Taper Forum

Location: Los Angeles, USA **Designer:** Rios Clementi Hale Studios **Completion date:** 2008 **Photos©:** Tom Bonner, Craig Schwartz

项目地点：美国，洛杉矶 设计师：里奥斯·克莱门蒂·霍尔工作室 完成时间：2008年 摄影师：汤姆·邦纳、克雷格·施瓦兹

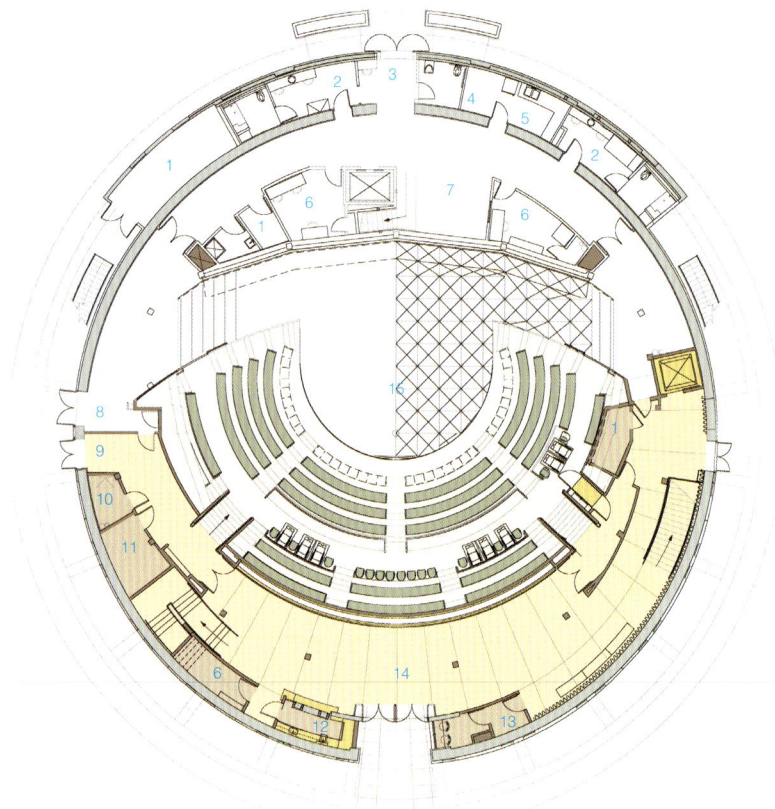

1. Storage	9. Exit	1. 仓库	9. 出口
2. Dressing room	10. Janitor closet	2. 更衣室	10. 门卫室
3. Back entrance	11. Mechanic & storage	3. 后门	11. 机械室和仓库
4. Restroom	12. Bar	4. 洗手间	12. 吧台
5. Break room	13. Ticketing	5. 休息室	13. 售票处
6. Office	14. Lobby	6. 办公室	14. 大堂
7. Green room	15. Stage	7. 演员休息室	15. 舞台
8. Loading entrance		8. 装卸入口	

One aim of the project was to achieve both functionality and preservation objectives. Rios Clementi Hale Studios and theatre consultant Sachs Morgan Studios worked together to respectfully update the building.

The architects fully enhanced the interior spaces with modern amenities for patrons and expanded the backstage areas for the actors and production teams. While maintaining the footprint of the original 1967 building, the architects cleverly carved out "found space" within the facility to better serve patrons and crew. From the newly configured entrance into the lobby that showcases an original Tony Duquette mural, to the auditorium with all-new seating, and the expansive lower-level lounge addition that provides patrons with larger restroom facilities, the architects successfully upgraded the space and added accommodations for disabled patrons.

The renovation allows for additional and more-spacious interior areas for the Taper's audience and artists. By reducing the ticket booth size, raising the lobby floor to be flush with the exterior ground plane, and moving the restrooms downstairs, the architects open up the lobby to present a more fitting entrance showcasing the original Tony Duquette abalone tile wall. The existing curved mosaic, whose mottled metallic green hues inspired much of the project's colour palette, is accented with new lighting – overhead and below – and a stainless-steel, flat-top guardrail to protect the abalone tiles while offering guests a place to set drinks during intermission. Extruded aluminum ceiling fins radiating from the abalone wall, along with integrated strip lighting, create a dynamic ceiling pattern. Terrazzo flooring reflects the tones and colours of the abalone shells, while continuing the mottled and shimmery look. With the more-spacious lobby, patrons are provided with easier access to a wet bar highlighted with zebra-wood cabinetry and varying-height black countertops.

The lounge is a major focus of the project. The new lounge is a comfortable, contemporary space. Rios Clementi Hale Studios gained additional square footage by utilising a portion of the underground parking garage to extend the theatre space, adding plentiful room for the 1,350-square-foot lounge. The broad curves of the walls reflect the overall shape of the building, giving the illusion that the lounge existed originally. The custom-designed, polished aluminum ceiling continues the circular motif and subtly recalls ripples of water in the reflecting ponds surrounding the circular Taper structure on the plaza level above. The circular flow of the restrooms directs patrons into the facilities, providing immediate access to the stalls, which cater to twice as many guests as before. Guests are then led back out to the lounge for a quick, in-and-out flow. A 12-foot-diameter marble centerpiece that functions as a hand-washing station on one side and a primping counter on the other side distinguishes the extra-spacious ladies' room. In the centre of the circular structure, rotating oval mirrors seem to blossom from the countertop. Each is affixed to a stainless-steel "stem" extending to the ceiling.

项目的目的之一是既保证建筑的功能性又保护其历史价值。里奥斯·克莱门蒂·霍尔工作室和剧院咨询专家萨克斯·摩根工作室共同合作，对建筑进行了升级改造。

建筑师运用现代设施全面提升了室内空间，也扩展了演员和制作团队所使用的后台区域。建筑师保留了1967年原始建筑的痕迹，在其内部打造了全新的空间。新建的入口通往挂着托尼·杜奎特壁画的大堂，礼堂里配备着全新的座椅，广阔的低顶休息室为观众提供了更大的洗手间。建筑师成功地对空间进行了升

1. The building is surrounded by greenery
2. A night view of the façade, brilliant and gorgeous

1. 建筑物周围绿树掩映
2. 建筑外立面夜景，流光溢彩

级，为残障人士提供了服务设施。

翻新工程为泰帕论坛的观众和艺术家们提供了更多、更宽敞的室内空间。设计师通过减小售票处的尺寸、抬高大堂的地面使其与室外地面齐平、将洗手间移到楼下等手段，将大堂打造成一个更合适的入口空间。原有的弧形马赛克瓷砖呈斑驳的金属绿色，这奠定了项目的基本色调。天花板和地面上的照明设施增强了色彩的表达。由不锈钢平顶护栏保护的鲍贝瓷砖区域为观众提供了喝一杯的场所。挤制铝天花板从鲍贝墙面上延展出来，上面装置着条形照明灯，形成了动感的天花板图案。水磨石地板倒映出鲍贝的色彩，延续了设计斑驳而闪光的特点。宽阔的大堂让观众可以简单地进入水吧，水吧以斑马纹木家具和高低不同的黑色台面为特色。

这个项目的主要焦点之一是休息区的设计。新的休息区是一个舒适、充满现代感的空间。设计者通过利用一部分地下车库增加了剧院的空间，为1350平方英尺的休息区增加了充足的空间。宽阔的弧形墙体反映出整栋建筑的形态，使这个新的休息区好像原来就存在似的。客户定制的、抛光的铝制天花板延续了圆形主题，巧妙地让人想到广场上环绕建筑外墙的水池。圆形、线条流畅的休息室直接引导剧院贵宾通向各个设施。客人们可以在休息室做临时性休息以及快速的往来出入。直径12英尺的大理石中央装饰品可以作为洗手台，也可以作为空间更大的女士休息室中的梳妆台。在这个圆形结构的中央，摆放了一圈椭圆形的镜子，好似花朵从中央绽放。每一面镜子都被固定在不锈钢的"花茎"上，一直延伸到天花板。

1. Restroom for women
2. Lounge
3. Storage
4. Family restroom
5. Restroom for men

1. 女士洗手间
2. 休息室
3. 仓库
4. 亲子洗手间
5. 男士洗手间

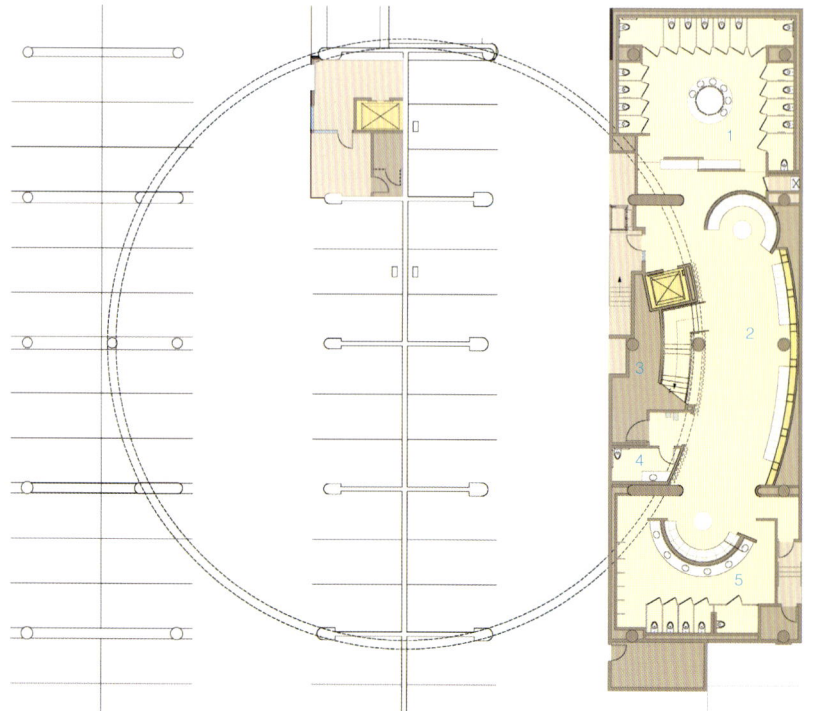

1. Main entrance
2. Entrance lobby
3. The existing curved mosaic, whose mottled metallic green hues inspired much of the project's colour palette

1. 主入口
2. 入口大堂
3. 原有的弧形马赛克瓷砖呈斑驳的金属绿色，奠定了项目的基本色调

1. Open and comfortable audience lounge
2,3. Entrance to the lounge
4. Restroom for women
5. Lobby bar

1. 宽敞舒适的观众休息室
2、3. 休息室入口
4. 女士卫生间
5. 大厅吧台

1,2. Auditorium

1、2. 礼堂

2

乌普萨拉音乐会及国会大厅 Uppsala Concert & Congress Hall

Location: Uppsala, Sweden **Designer:** Henning Larsen Architects **Completion date:** 2007 **Photos©:** Henning Larsen Architects

项目地点：瑞典，乌普萨拉 设计师：亨宁·拉森建筑公司 完成时间：2007年 摄影师：亨宁·拉森建筑公司

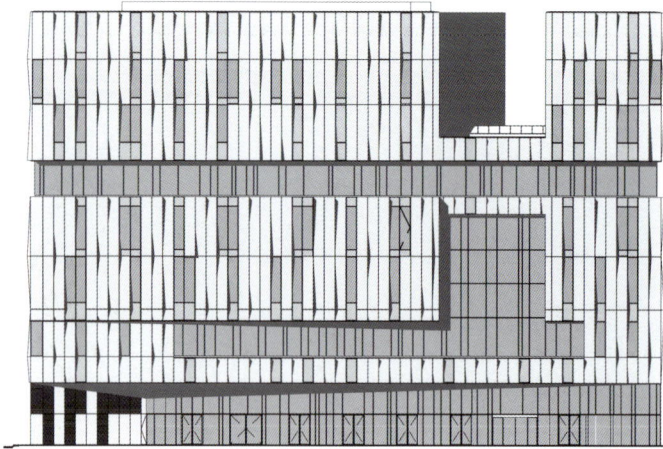

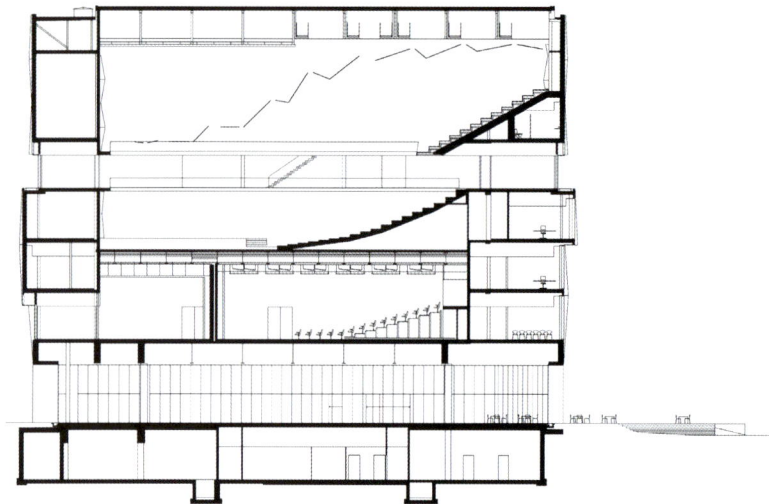

On the arrival level (Level 2) an exhibition and banquet hall that holds up to 600 people is located, as well as a public café with a terrace that faces the south. The long red escalators connect various foyer areas leading the visitors from the arrival hall to Level 3, where they are met by a magnificent view of Vaksala Square. In connection to the foyer, Level 3 comprises a hall with room for 350 people for conference and music and a hall for 100 people for larger meetings. Covering the length of three floors the escalators continue from Level 3 to the large hall's great foyer on Level 6 that offers a unique panoramic view of Uppsala's rooftops and historical skyline.

The large hall is the most important room in the house. It seats 1,150 people from stalls to balcony. With its sublime acoustics and technical flexibility, the hall is designed for a large and versatile repertoire covering everything from symphonic concerts to modern pop and jazz. Level 3, 4 and 5 mainly comprise offices, rehearsal halls, storage rooms, boxes, stage entrance and backstage areas. Additionally, Level 7 comprises an independent conference hall. The new concert and congress hall is a modern and dynamic house. It has its own impressive characteristics but is designed with respect to and in interplay with the other significant buildings in Uppsala.

二层（入口层）设有展览厅及宴会厅以及南向咖啡厅，展览厅及宴会厅能够同时容纳600人。长长的红色扶梯将门厅的不同区域有机联系在一起，引领人们直达三楼，将瓦克莎拉广场美景尽收眼底。三楼设有能够容纳350人的会议和音乐厅以及百人大型会议厅。沿扶梯直上，人们即可到达建筑的顶层，尽情饱览乌普萨拉城市风光。

大厅是整个建筑的亮点，能够同时容纳1150人，其完美的音响效果和科技应用能够承办各种大型交响音乐会、现代流行音乐和爵士乐欣赏会。3楼、4楼和5楼设有办公空间、排练厅、储藏室、舞台入口及后台。7楼设有独立的会议厅。新音乐会和会议厅将现代和动感的设计理念完美诠释，强调个性化设计的同时，与乌普萨拉市的其他建筑完美融合得恰到好处。

1. Distant view of the building   1. 建筑远景
2. Main façade   2. 建筑外观
3. Main façade night view   3. 建筑外观夜景

2

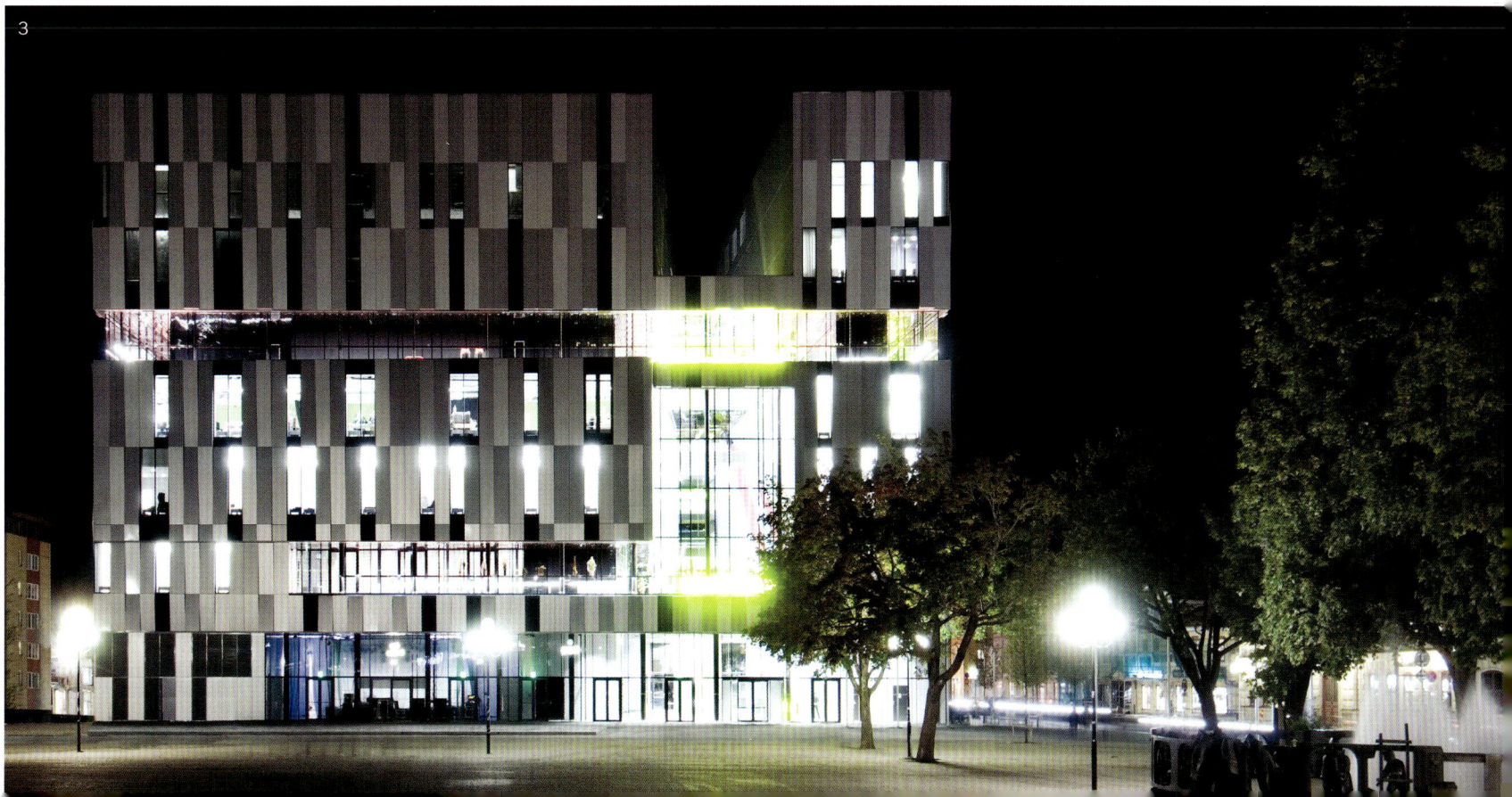

3

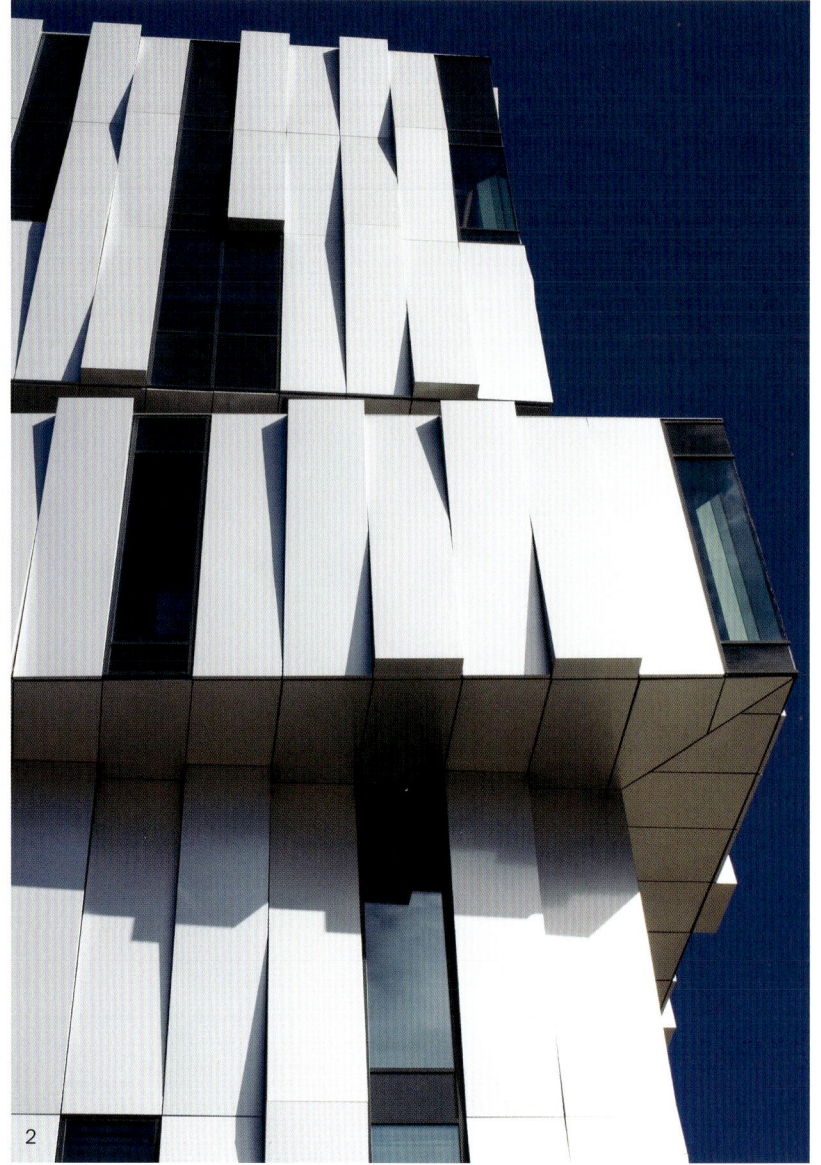

1, 2. Façade detail
3. View showing upper foyer and entrance on the ground floor

1、2. 外观细部
3. 上层大厅和一楼入口

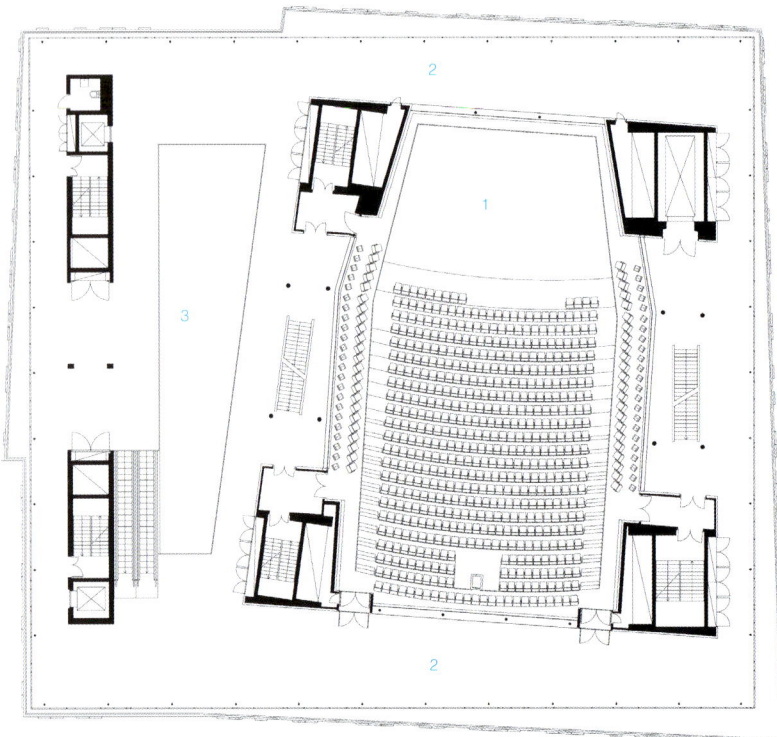

1. Stage
2. Corridor
3. Lounge

1. 舞台
2. 走廊
3. 休息室

1

2

1, 2. Audiences can enjoy good views through the glass windows at foyer
3. The large hall

1、2. 观众可以通过大厅的玻璃窗看到外面的景色
3. 大厅

里摩日音乐厅 Limoges Concert Hall

Location: Limoges, France **Designer:** Bernard Tschumi Architects **Completion date:** 2007 **Photos©:** Christian Richter **Award:** AIA New York State Design Award, 2007

项目地点：法国，里摩日 设计师：伯纳德·曲米建筑事务所 完成时间：2007年 摄影师：克里斯汀·里克特 所获奖项：2007年美国建筑师协会纽约分会设计奖

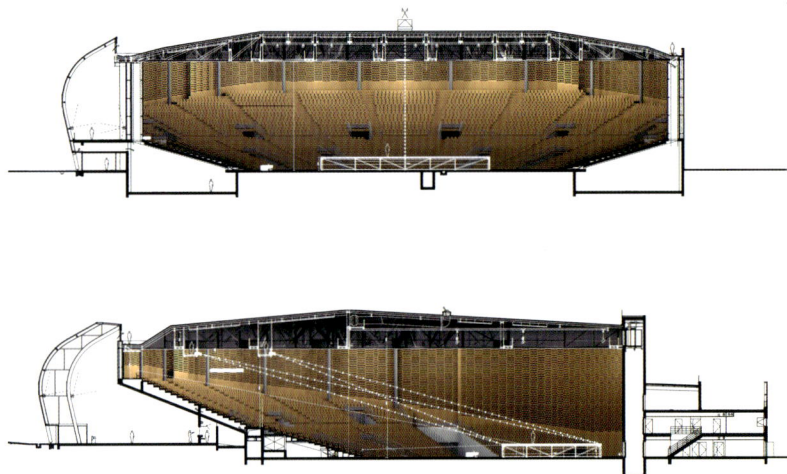

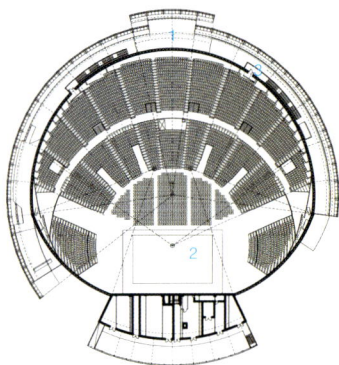

The Concert Hall in Limoges, located in the centre of France, returns to the general envelope concept already explored in Rouen, but transforms it through a new material strategy. If architecture is the materialisation of a concept, what if the concept remains the same, but the material changes? The designers decided to explore the implications of such a transformation with a new variation on a familiar programme. In Rouen, the outer envelope was made of steel and the inner envelope of exposed concrete. In Limoges, the outer envelope is made of wood arcs and translucent rigid polycarbonate sheets and the inner envelope of wood.

The use of wood was suggested by the location of the hall, in a clearing within a large forest surrounded by trees over 200 years old. The region also has an active timber industry. In addition, the soft translucency of the polycarbonate complemented the wooden frame by allowing light to filter in and out of the building. The strategy establishes reciprocity between concept and context.

The configuration of the double envelope with circulation in-between is a scheme that is advantageous for both acoustical and thermal reasons. In Limoges, the designers modified the basic typology to respond to several siting issues: whereas Rouen's spiral aimed at channeling the lateral movement of crowds entering the building sideways, Limoges's detached and fragmented envelope opens in two directions, towards both the forest and the road. Between the two envelopes are the movement vectors: two ramps, one extending downward toward the lower tiers of the auditorium, and the other upward toward the upper tiers. Additionally, two straight "flying" staircases extend directly toward the top row of seats.

Much of the material treatment is determined by energy conservation and sustainability considerations. The 5-centimetre-thick semi-rigid polycarbonate sheeting, with its multiple inner layers of cells, provides excellent insulation value. The highest portion of the façade has a pixelated design silk-screened directly on the shell for additional solar protection.

Natural ventilation is integrated into the concept, so that the climate of the foyer can be kept at a temperate level, with little additional heating required. Acoustics play a major role in the treatment of the inner envelope, both internally and externally. In the auditorium, strong absorption is required for an 8,000-spectator-capacity hall, while in the large, 1,800-square-metre foyer, absorbent and reflective materials are alternated to generate more varied ambiences and acoustic effects.

音乐厅位于法国中部的里摩日，遵循了鲁昂音乐厅的基本设计理念，但是采用了全新的材料策略。如果说建筑是概念的物化，那么如果保持同样的概念，只是改变材料会怎么样？设计师决定对此进行尝试。在鲁昂，建筑的外层表皮由钢铁制成，内层表皮由露石混凝土制成；在里摩日，外层表皮由弧形木条和透明的聚碳酸酯板制成，内层表皮由木材制成。

木材的使用由音乐厅所在的场地决定，它位于一片平均树龄超过200年的森林的空地之上。该地区拥有十分活跃的木材工业。此外，补充木框架的聚碳酸酯板柔和的透明感保证了光线的进出。整个设计是概念和环境相互作用的结果。

双层表皮结构和中间走道对音效和隔热都十分有利。设计师在里摩日音乐厅的设计中考虑了场地元素。鲁昂的螺旋结构旨在保证人流的横向行动；而里摩日的分离表皮则向两个方向开口，一面朝向森林，一面朝向公路。双层表皮之间是移动空间：两个坡道、一个通往下层礼堂，一个通往上层。此外，两个直线型楼梯直接通往上层的坐席。

许多材料的处理都出于节能和可持续性的考虑。5厘米厚的半刚性聚碳酸酯板拥有多层内部单元格，具有优秀的绝缘性能。外墙的最上部有一个像素化设计的丝印，提供了附加的日光保护措施。

设计采用了自然通风，因此，大厅能保持恒温，几乎不需要附加加热系统。内层表皮在音响效果上起到了重要作用。可容纳8,000人的礼堂需要有力的隔音设备；1,800平方米的大厅中的吸音和反射材料形成了多变的氛围和音响效果。

1. Lobby
2. Stage
3. Entrance
1. 大堂
2. 舞台
3. 入口

1. Night view of main façade
2. Façade detail and entrance
1. 建筑外观夜景
2. 外观细部和入口

1. Overall view of exterior
2. Dusk façade view

1. 建筑外观全景
2. 黄昏中的建筑

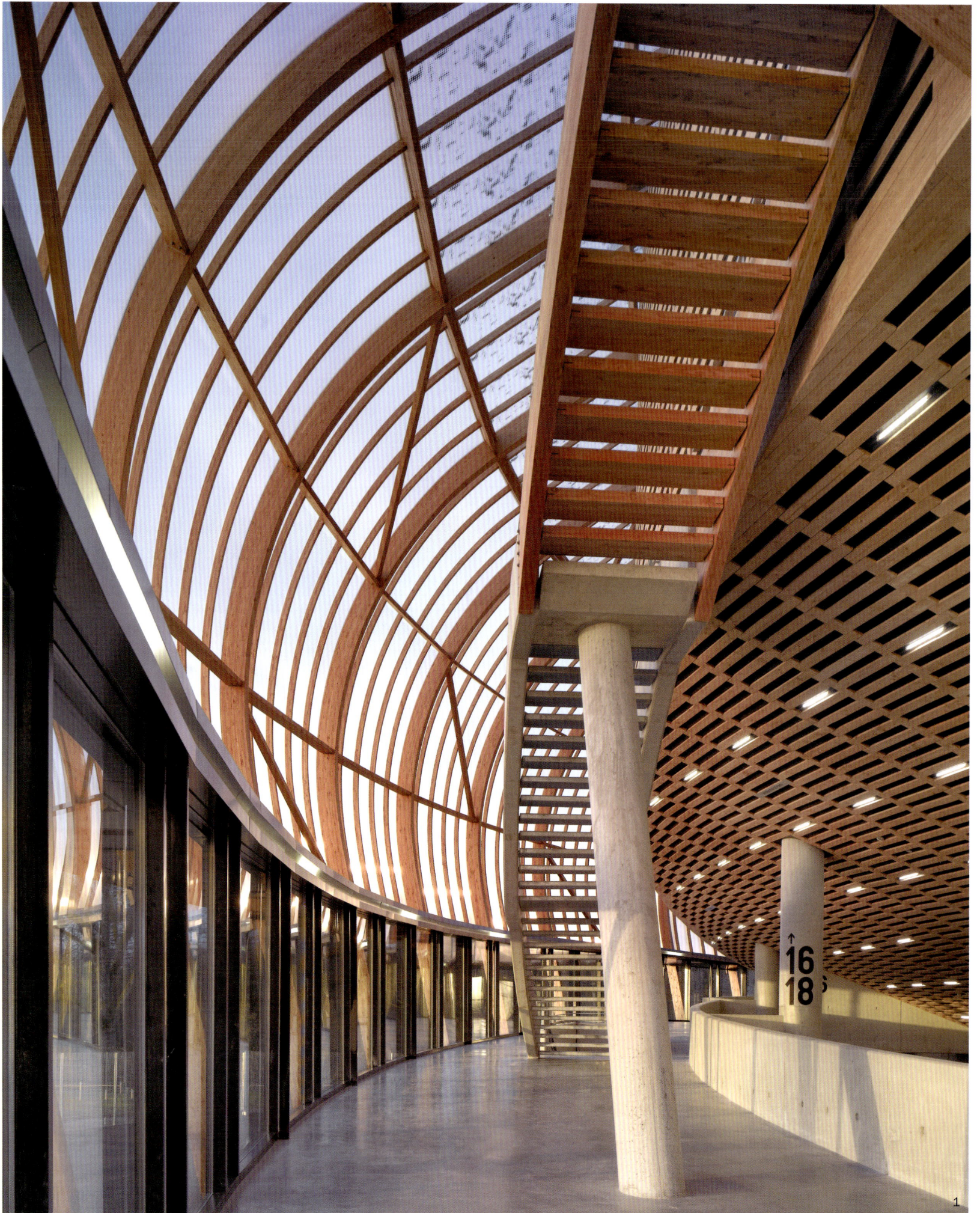

2

1. Roof and curtain wall detail
2. Lobby

1. 屋顶和幕墙细部
2. 大堂

The Concert Hall, Aarhus, Extension

奥尔胡斯音乐厅扩建工程

Location: Aarhus, Denmark **Designer:** C. F. Møller Architects **Completion date:** 2007 **Photos©:** Courtesy of C. F. Møller Architects **Area:** 17,400 square metres

项目地点：丹麦，奥尔胡斯 设计师：C·F·穆勒建筑事务所 完成时间：2007年 摄影：C·F·穆勒建筑事务所提供 面积：17,400平方米

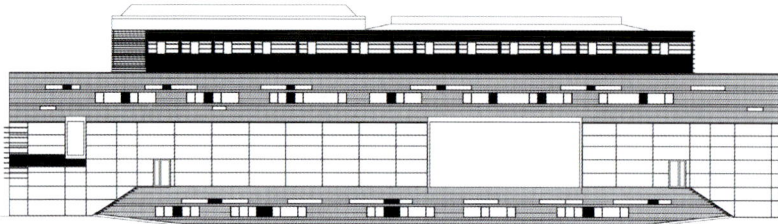

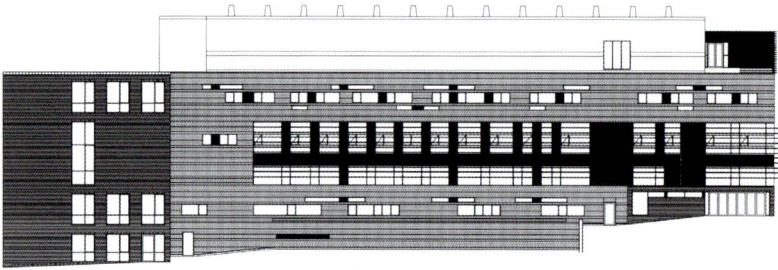

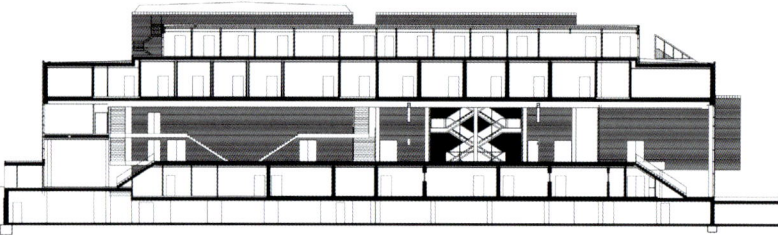

The Concert Hall, Aarhus was originally designed by Kjær & Richter, and was inaugurated in 1982. With its new extension the Concert Hall has doubled in size and now encompasses a wide range of functions that turn the total complex into a unique concert and educational institution of international standard. The extension includes a Symphonic Hall with sublime acoustics and room for an audience of 1,200, a Rhythmic Music Hall, a Chamber Music Hall, and premises for the Royal Academy of Music, Aarhus, the Aarhus Symphony Orchestra, the Filuren Children's Theatre Ensemble and the Danish National Opera.

The Symphonic Hall is specially built for symphonic music, and functions in itself like a giant instrument, the sound of which can be adjusted by moving walls, carpeting and acoustic panels. The Hall's internal dimensions are carefully designed to produce the best acoustics; the model for its proportions is the Golden Hall of the Musikverein, Vienna, which is optimum for orchestral music. Despite its world-class performance, the hall was built with modest means, using mainly pre-fabricated concrete panels.

The colours in the Hall are Nordic: light ash wood, silver-grey concrete panels and black seating. The acoustic panels are cladded in red fabric, and have both an acoustic and an architectural function: the powerful red colour means the panels can clearly be seen when they are extended and thereby adapted to the music, and in this way, they help to say something about the music's intensity.

C. F. Møller Design has been responsible for a new reception, furniture design for the lobby area, the construction of new kitchen and canteen with a bespoke buffet design, and general signage. The furnishing includes the especially developed Buco lobby series, with benches and foyer tables with a distinct cutout that allows the architecture stand out, and accommodates handbags etc.

Everything is the same material and colour scheme, simple and bright with the use of white composite materials to create an architectural coherence and contrast to the characteristic yellow brick walls and artist Ingvar Cronhammer's distinctive contribution to the building, including 17 large LED chandeliers and a red laser-illuminated water feature.

奥尔胡斯音乐厅开幕于1982年，由科加尔&里克特设计。音乐厅的扩建工程使其面积扩大了一倍，将建筑变成了独特的音乐会和教学设施的综合体。扩建工程包括一个可容纳1,200人的交响乐厅、一个节奏音乐厅、一个室内音乐厅、以及皇家音乐学院、奥尔胡斯交响乐团、菲洛仁儿童剧院和丹麦国家歌剧团的基础设施。

交响乐厅特别为演奏交响乐而建，自身就像一件巨大的乐器。墙壁、地毯和隔音板的移动都可以调节音响效果。音乐厅的内部尺寸经过了精心设计，以达到最佳的音响效果。它的设计原型是维也纳金色大厅。尽管音乐厅具有世界级的音响效果，它所采用的建筑材料十分简单，主要为预制混凝土板。

音乐厅的色彩具有北欧特色：浅灰色木材、银灰色混凝土板和黑色座椅。隔音板包裹着红色布料，兼具声学和建筑功能：强烈的红色让隔音板在拓展的时候十分明显，也显示出音乐的强度。

C·F·穆勒建筑事务所设计了一个新接待处，进行了大堂区域的家具设计、新厨房和自助餐厅的设计和总体引导标示设计。家具包括特别开发的巴克大堂系列，长椅和桌子的独特设计让建筑更加突出，还能容纳手提包等物品。

建筑采用了相同的材料和色调搭配，白色的复合材料简洁而明快，营造出建筑的一致性，与突出的黄色砖墙、英格瓦·克罗恩翰墨设计的17个巨型LED吊灯和红色激光照明水景形成了鲜明对比。

1. Open-air cafe　1. 露天咖啡厅
2. Facade detail　2. 外观细部

1. Main facade and side entrance
2. Lobby/reception

1. 建筑外观和入口
2. 大堂/前台

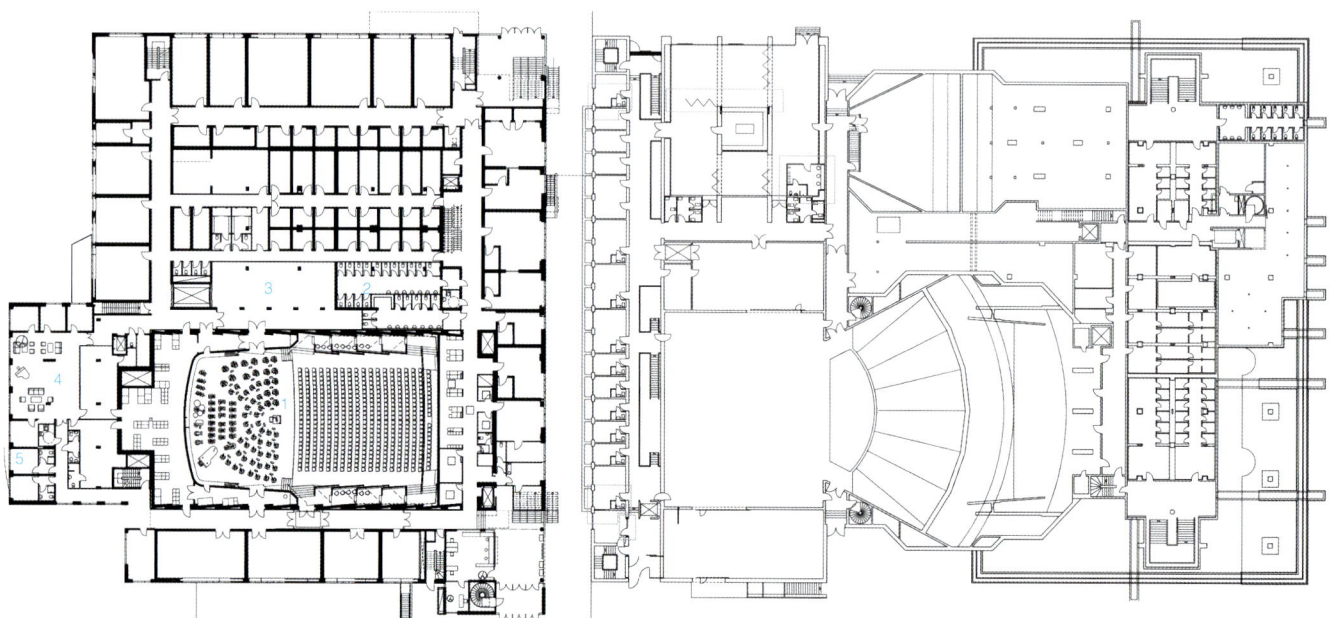

1. Concert hall
2. Toilets
3. Foyer
4. Lounge
5. VIP restroom

1. 音乐厅
2. 洗手间
3. 大厅
4. 休息室
5. VIP休息室

1-3. Concert hall
4. Mezzanine/VIP seat

1-3. 音乐厅
4. 夹层/VIP坐席

地方现代音乐中心

Regional Centre of Contemporary Music

Location: Nancy, France **Designer:** Périphérique Architectes/Anne-Françoise Ju meau, Emmanuelle Marin, David Trottin **Completion date:** 2007 **Photos©:** Luc Boegly

项目地点：法国，南锡 设计师：佩里菲力克建筑事务所/安妮·弗朗西斯·朱莫、艾曼纽·马林、大卫·特洛汀 完成时间：2007年 摄影师：卢克·伯格利

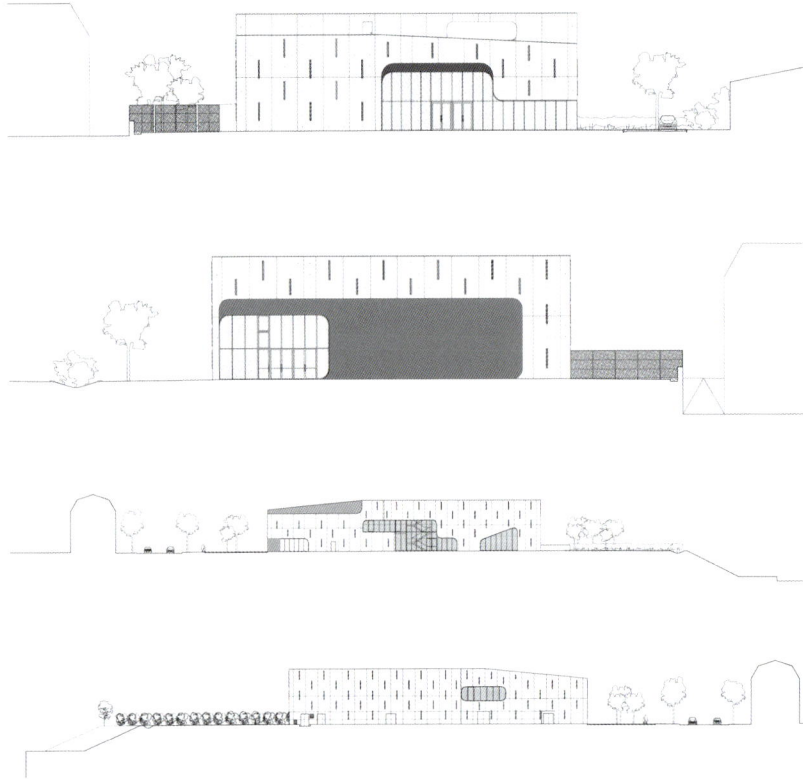

The project for the Regional Centre of Contemporary Music participates in the overall urban plan for green spaces in this huge district of Nancy, which is in the process of radical transformation. Within this dynamic setting, the aesthetics of the new building embodies and reinforces the innovative aspect of the site by the diffusion, production, and fabrication of a contemporary musical culture.

Architects chose to set the facility in the exact area occupied by the lot while scrupulously respecting the alignments and overall dimensions of the surrounding buildings. The architectural principle induced by this respectful urban positioning was to design a building that is both highly compact and widely open to its surroundings.

The principle is to "hollow out" this urban space, characterised by its powerful, compact architectural style, and containing the very core of the programme (the concert halls, the studios, offices, dressing rooms, etc.), and to pierce through it an interior street in bright red, situated on the ground level to facilitate access for the public and for equipment. This interior street is the user-friendly and functional "Fil Rouge" or "Connecting thread" (literally, "red thread") of the project: it reveals and leads to the entire interiority of the facility.

The building is a solid block of concrete, slightly red tinted by the metal shuttering panels used to cast it. The texture of all of the façades is animated by a pattern of cuts in the walls, housing recessed lights that liven up the building differently during the day and at night when it can be turned into a magic lantern in the city, clearly identifying the building as a music centre.

The main entrance to the RCCM is located on Boulevard d'Austrasie. It is designed in the manner of a large covered porch on the scale of the boulevard, which affords a view of the depths of the facility, straight through to the Bras Vert, or Green Arm. The extensively glass-enclosed reception area is placed at the angle of this urban window. This reception area, with its high inclined ceiling, leads to the small concert hall (250 seat) through the double-entrance security door integrated into a service and utility area, including the DJ booth for concerts in the hall, the ticket window and coat-check, and merchandise stands. The second part of the "Fil Rouge" is the Walkway. It leads to the multimedia areas and the big concert hall (1,300 seats) and serves as an exhibition space. On the "Bras Vert" side, the "Fil Rouge" ends in a big bay window that opens onto the canal.

地方现代音乐中心项目是南锡地区绿色空间整体规划的一部分，整个地区正经历着巨变。在这样活跃的背景之下，新的建筑通过现代音乐文化的传播、创作和加工增添了场地的创新感。

建筑师选择将音乐中心打造成一个紧凑的空间，并且与周边建筑的规模和区域总体规划相适应。因此，建筑既结构紧凑，又向周边地区广泛开放。

项目的设计原则是"挖空"城市空间，以有力、简洁的建筑形式来承载着功能空间（如音乐厅、工作室、办公室、更衣室等）。一楼，一条明快的红色室内街道横穿建筑，形成了主要公共路径。市内街道宛如一条红线，将整个室内空间贯穿起来。

建筑是一个坚固的混凝土方块，上面装饰有红色的金属板。雕刻图样为建筑外墙增添了质感，嵌入式灯光让建筑在白天和夜晚有不同的感觉。晚上，建筑会变成城市里的魔术灯，让人一眼就看出这是音乐中心。

音乐中心的主入口位于奥斯特拉斯大道上，宛如大道上一个巨大的门廊，可以直接通入建筑深处的"绿色手臂"区域。宽阔的玻璃接待处就紧邻这个城市窗口。拥有高大天花板的接待处通过服务区的双入口安全门连接小音乐厅（可容纳250人），其间会经过DJ室、售票窗口、衣帽检查处和商业柜台。"红线"的另一部分是走道，它通往多功能区和大音乐厅（可容纳1,300人），是一个展示空间。在"绿色手臂"的一侧，"红线"以一个巨大的凸窗为重点，遥望着运河。

1. Overall view of exterior   1. 建筑外观全景
2. Entrance   2. 入口
3. Dusk view of main façade   3. 黄昏中的建筑

2

3

1. Lobby
2. Corner of lobby
3. Lobby/resting area

1. 大堂
2. 大堂一角
3. 大堂/休息区

3

1. Small salle
2. Loge
3. Studio 1
4. Studio 2
5. Catering
6. Administration
7. Space associations
8. Professional multimedia
9. Studio 3
10. Studio 4
11. Big salle

1. 小表演厅
2. 包厢
3. 工作室1
4. 工作室2
5. 餐厅
6. 行政区
7. 空间协会
8. 专业多媒体室
9. 工作室3
10. 工作室4
11. 大表演厅

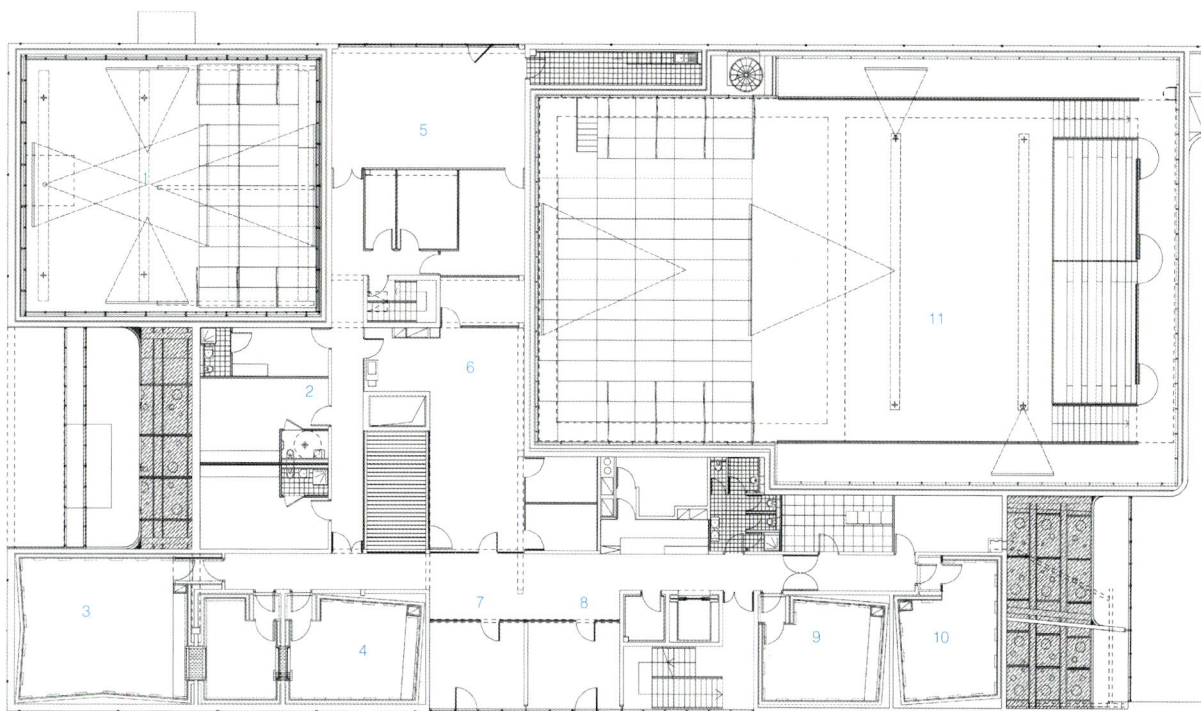

2

1. Office
2. Hallway

1. 办公室
2. 走廊

1, 2. Concert hall
3. Office
4. Studio

1、2. 音乐厅
3. 办公室
4. 工作室

21号馆迷你歌剧院 Pavilion 21 Mini Opera Space

Location: Munich, Germany **Designer:** Coop Himmelb(l)au **Completion date:** 2010 **Photos©:** Duccio Malagamba **Footprint:** 560 square metres

项目地点：德国，慕尼黑 设计师：库伯·西梅布芬 完成时间：2010年 摄影师：杜契尔·马拉甘巴 使用面积：560平方米

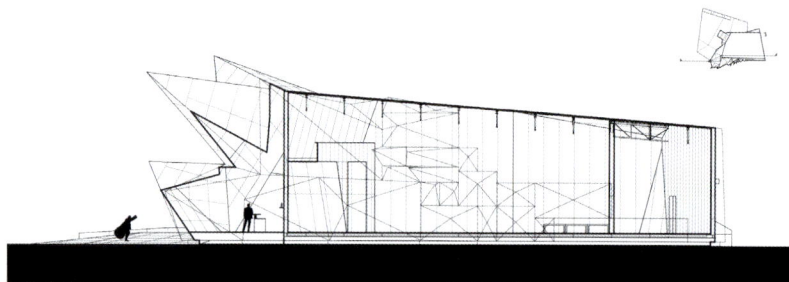

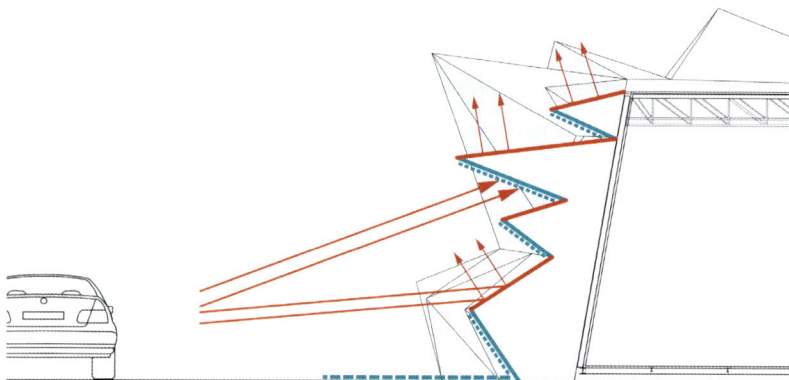

The task was to create a space with 300 seats (or 700 standing spectators) for experimental performances of the Bavarian State Opera. The Pavilion should be dismountable, transportable and re-mountable and make the respective urban space distinctive through its shape.

Mass and therefore weight are the decisive criteria for good acoustics. The conception of the Pavilion 21 Mini Opera Space therefore had to overcome a contradiction: to design a lightweight construction, which must allow being disassembled and re-assembled quickly, but which at the same time meets the acoustical requirements of a concert hall.

The idea to combine architecture with music is not new. Also the term soundscaping is not new. Similar to landscaping it involves "Gestalt". Soundscaping originates in the 1940s and designates a method of composing. The strategy to achieve soundscaping comprises three steps: firstly, to realise the shielding effect between square and street, secondly, to shape the geometry of the Pavilion in such a way that the surface deflects noise, and thirdly, to design the surface of the Pavilion in such a way that it reflects and absorbs sound.

As a starting point towards the abstraction of music into spatial form, a sequence from the song "Purple Haze" by Jimi Hendrix and a passage from "Don Giovanni" by Mozart were transcribed. Through the analysis of frequence sections from these pieces of music and through the combination with the computer generated 3D model, the sequences are translated into pyramidal "spike constructions" by means of parametric "scripting".

In order to implement the objectives of the interior spatial acoustics, the interior wall and ceiling surfaces were fitted with a combination of perforated absorbing and smooth reflecting sandwich panels. The flooring of the Pavilion is carried out as a reflecting even "stage floor". Sound reflecting, parallel wall and ceiling surfaces are avoided and are therefore tilted or skewed.

项目的任务是为巴伐利亚国家歌剧院打造一个300坐席（或700个站席）的试验性演出空间。场馆要具有可拆卸性、可移动性和可重装性，还要有抢眼的造型。

规模和重量是音响效果的决定性因素，因此，21号馆迷你歌剧院必须克服一个矛盾：其结构必须足够轻盈，以适应拆卸和组装，同时，还要保证音乐厅的音响效果。

将建筑与音乐结合在一起，或称其是音乐景观设计并不是什么新概念。与地面景观设计相似，它有形态上的要求。打造音乐景观有三个步骤：一是实现广场与街道之间的屏蔽效果；二是打造场馆的几何造型，让表皮隔离噪音；三是让场馆的表皮反射和吸收声音。

作为将音乐抽象成空间形态的起点，设计师转录了吉米·亨德里克斯的歌曲《紫色烟雾》和莫扎特的《唐乔凡尼》的选段。通过对这些音乐作品的研究和电脑3D建模，场馆被打造成尖锐而凸出的锥体结构。

为了保证室内音响效果，室内的墙壁和天花板上安装了穿孔吸音和平滑反射夹心板。场馆的地面是一个反射的舞台。声音反射、平行墙壁和天花板表面都被有效地避开了，所有的表面都是倾斜或扭曲的。

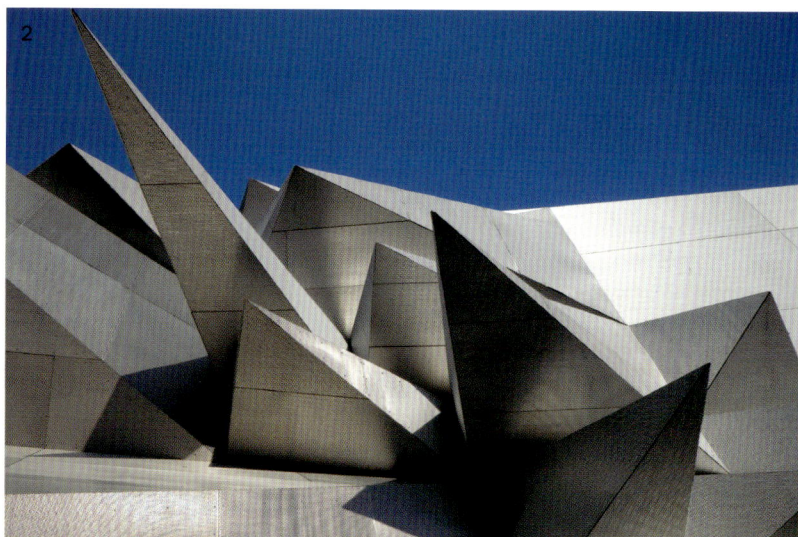

1. Façade
2. Façade detail
3. Overall view of the façade, surrounded by other buildings
4. Entrance

1. 建筑外观
2. 外观细部
3. 建筑外观全景，周围是其他建筑
4. 入口

3

4

1

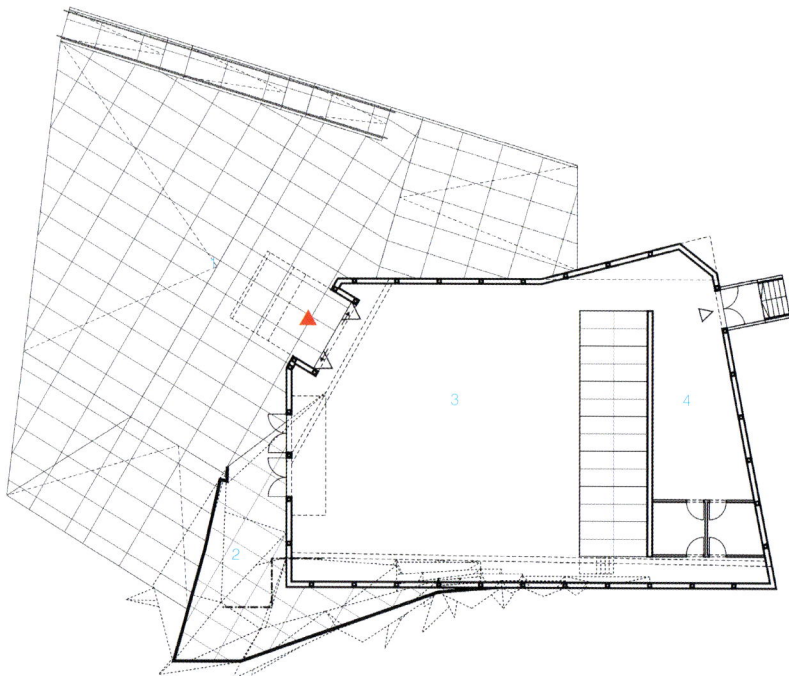

1. Platform
2. Lounge/bar
3. Auditorium for 300 persons
4. Backstage

1. 讲台
2. 休息室/吧台
3. 可容纳300人的礼堂
4. 后台

1. View showing the opera space from upper level
2. Distinctive shape in dusk
3. Shape detail

1. 从上方看歌剧院
2. 黄昏中，建筑独特的造型
3. 造型细部

1. Night view, performance is beginning
2. Night view of entrance plaza
3. Façade detail at night

1. 建筑夜景，演出正要开始
2. 入口广场夜景
3. 夜晚建筑细部

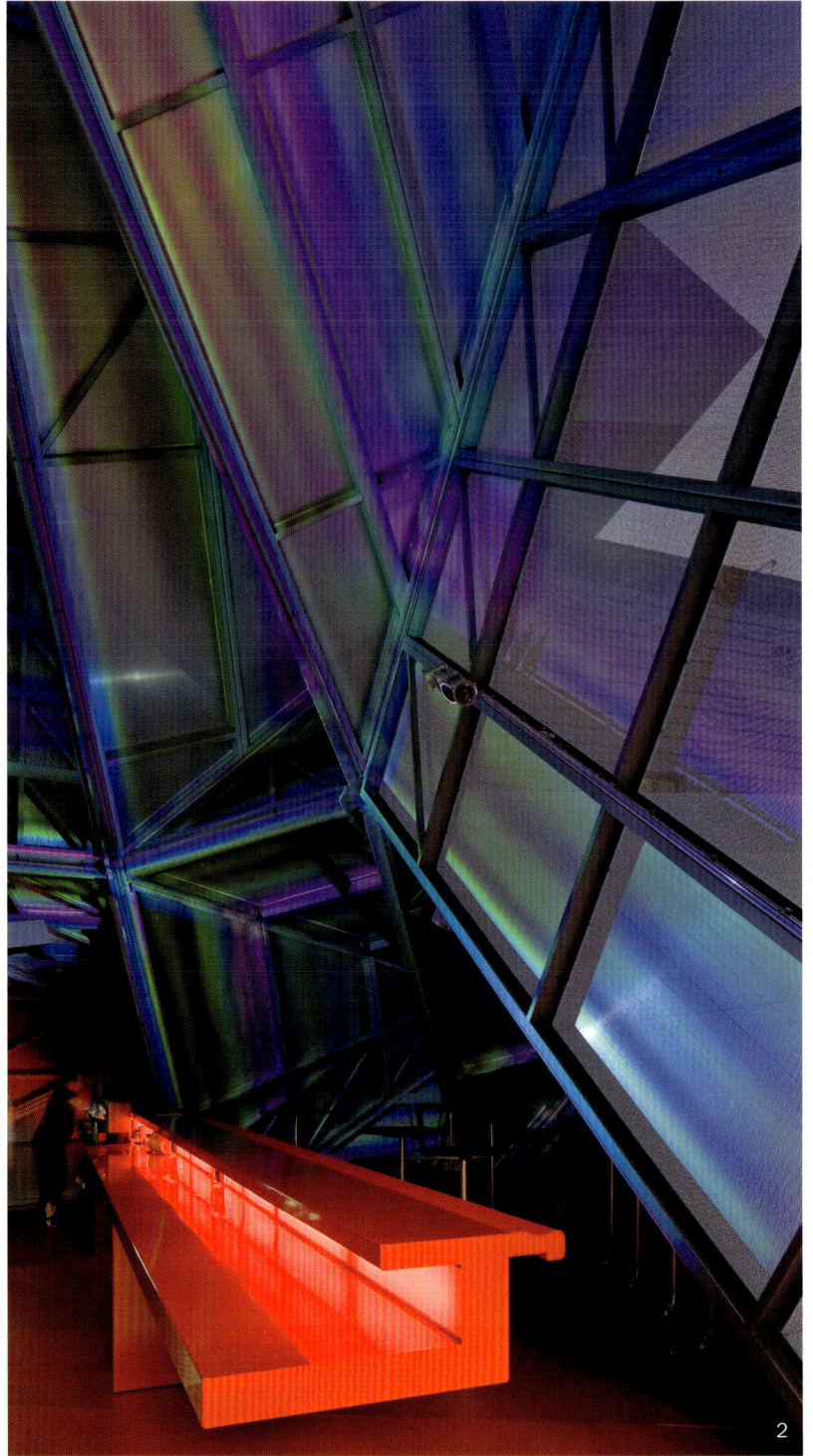

1. Drink bar
2. Resting area
3. Platform
4. Rehearsal space

1. 饮品吧
2. 休息区
3. 讲台
4. 排练空间

萨弗拉城市剧院 # Municipal Theatre Of Zafra

Location: Zafra, Spain **Designer:** Enrique Krahe/Aranzazu Montero, C. Brage, E. Espinosa, L. Fernandez, J. Isla, J. Longhi, D. Pérez **Completion date:** 2009 **Photos©:** Miguel de Guzmán **Area:** 4,408 square metres **Awards:** LAMP Lighting Award 2010/IALD Award of Mention 2011

项目地点：西班牙，萨弗拉 设计师：恩里克·卡拉和/阿兰扎祖·蒙特罗、C·布拉吉、E·埃斯皮诺萨、L·费尔南德斯、J·伊斯拉、J·隆基、D·佩雷斯 完成时间：2009年 摄影师：米格尔·德·古兹曼 面积：4,408平方米 所获奖项：2010年LAMP灯光设计奖；2011年国际灯光设计师协会提名奖

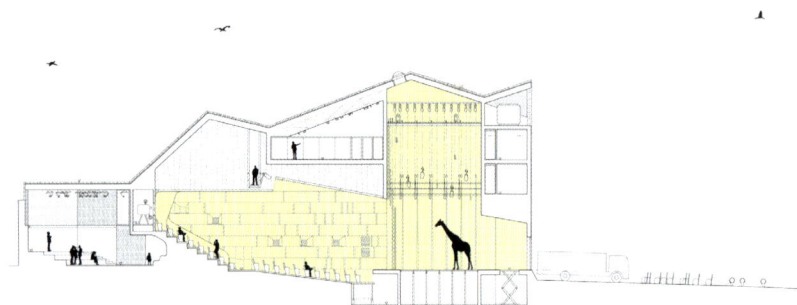

Located in a small town in the Southwest of Spain, the Municipal Theatre of Zafra aims to reconcile the different events that are everyday staged, solving the transition between the monumental city and a "periphery" under construction, through a smooth connection with the pedestrian centre.

The new organisation provides free paths with gentle slopes that allow full accessibility for visitors and restricted to vehicles. Cobblestone pavements and terraced gardens delimit the perimeter of the plot, consisting of old buildings in two of which vernacular elements (masonry vaults and slate walls) were consolidated in order to preserve them for a future enlargement of the cultural activities.

The design tackles the desirable versatility of this type of facilities in a small city (film, theatre, concert hall...), with the inevitable technical adequacy. But beyond a mere functional or contemplative affection, the building challenges the traditional role of spectators as passive ingredients, inviting them to achieve certain complicity with both visitors and passers-by, so they can get involved not only when a representation is taking place: for instance, the stage (in addition to offering a position that favours the direct loading) can literally open to the city, while the seats of the orchestra are coloured pixels that, viewed from the stage when empty, compose the anamorphosis of an always-looking eye surrounded by the natural felt finished walls of the main space.

Theatres are supposed to be venues functionally designed to stimulate some sort of reverie. Therefore, they have traditionally restricted and moderated the effect of natural light, in favour of a sophisticated lighting control, which rarely is fed back by its typological or programmatic particularities. At the Theatre of Zafra, lighting becomes a main issue, and shares the general concepts underlying the project. Thus, it acquires an active responsibility that reinforces the playful nature of the programme, creating a second graphic level, guiding visitors through the less crowded areas, and helping to create a calm and evocative atmosphere, as an accomplice preparation for the drama representation that awaits.

萨弗拉剧院位于西班牙西南部的一个小镇，旨在安排城市中各种类型的日常演出，通过与步行街中心的平稳连接，成为城市与城市外围的过渡区。

新项目的走道带有缓坡，可以让行人和限制车辆出入。卵石路面和平台花园划分了场地的界限。这一区域的建筑拥有两个鲜明的本地元素——砖石拱顶和板岩墙壁。建筑采用了这两种元素，以便未来进行文化活动的扩建。

设计通过各种技术手段融合了这一类型设施的各种功能（例如电影院、剧院、音乐厅……）。建筑不仅具有功能性和深思性，还挑战了观众传统的被动角色，邀请他们与游客和行人形成互动，使他们参与到活动中来。例如，舞台可以向城市开放，乐队的座椅是彩色的，当座椅空着时，从舞台向下看，整个观众席看起来像一只睁开的眼睛，与空间的墙面装饰自然地结合在一起。

剧院的功能本应该激起人们的沉思。因此，剧院深受自然光线的限制和调控，或受到精密的灯光控制，极少受到自身的类型和项目特性所影响。在萨弗拉剧院，照明成为了主要元素，是项目的根本理念之一。此外，剧院还充分强调了项目的趣味性，通过次要的形象楼层，指引游客前往不拥挤的区域，便于营造出一个冷静而具有情绪感的氛围，为即将上演的戏剧做了准备。

1. Terrace detail
2. General day view of the theatre
3. Night view

1. 平台细部
2. 剧院日间全景
3. 夜景

2

3

2

1. Interior hallway
2. Foyer and lounge

1. 室内走廊
2. 门厅和休息室

Ground floor plan

1. Stage space access
2. Main access
3. Actors' access
4. Multi-use building
5. Auxiliary building
6. Electrical building
7. Existing building
8. Cistern
9. Patio
10. Loading/unloading
11. Foyer
12. Auditorium
13. Box office
14. Cloakroom
15. Rest room
16. Rest room
17. Adapted rest room
18. Stage
19. Forestage
20. Backstage
21. Lift
22. Dimmers/rest room
23. Utility room
24. Ventilation room
25. Power generator

一层平面图

1. 舞台空间入口
2. 主入口
3. 演员入口
4. 多功能楼
5. 附属楼
6. 电力楼
7. 原有建筑
8. 水池
9. 露台
10. 装卸区
11. 门厅
12. 礼堂
13. 售票处
14. 衣帽间
15. 洗手间
16. 洗手间
17. 改造的洗手间
18. 舞台
19. 前台
20. 后台
21. 电梯
22. 调光器/洗手间
23. 杂物间
24. 通风室
25. 电力发电机

1. Foyer to the terrace
2. Water closet
3, 4. Theatre hall

1. 平台门厅
2. 洗手间
3、4. 剧院大厅

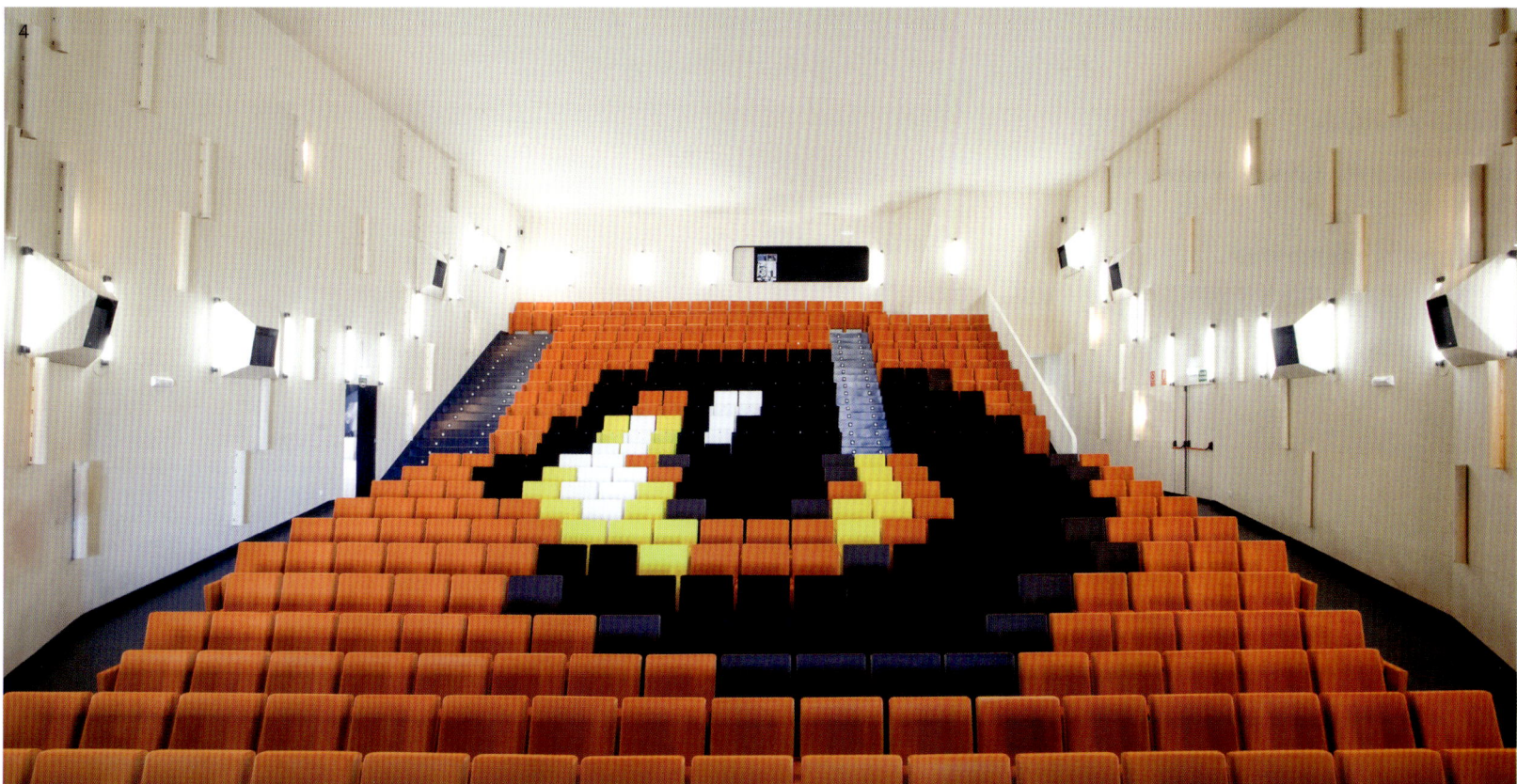

卡特索斯剧院 # Théâtre de Quat'sous

Location: Montreal, Canada **Designer:** Les Architectes FABG **Completion date:** 2009 **Photos©:** Steve Montpetit **Area:** 860 square metres **Awards:** 2010 USITT Architecture Merit Award

项目地点：加拿大，蒙特利尔 设计师：FABG建筑事务所 完成时间：2009年 摄影师：史蒂夫·蒙特波蒂特 面积：860平方米 所获奖项：2010年美国剧院技术协会建筑荣誉奖

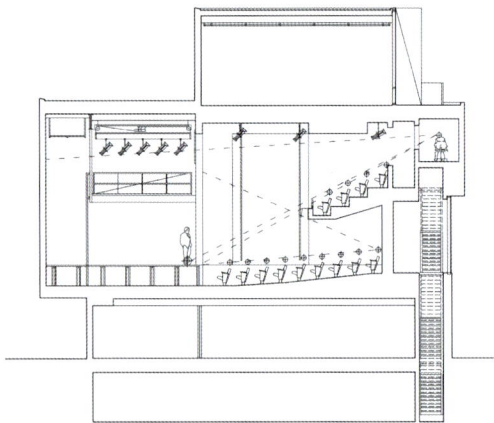

Théâtre de Quat'sous was established in 1963 by a group of actors under the direction of Paul Buissonneau. Together they bought a synagogue located on Avenue des Pins in Montréal and transformed it into a small theatre that never ceased to present daring productions from young creators including the mythical psychedelic happening "Osstidcho" in May 1968.

Serious problems related to the security and comfort of the users combined with the need for production spaces led to the decision to demolish and rebuild Théâtre de Quat'sous on its actual site after many feasibility studies demonstrated the impossibility to renovate the crumbling wood and brick structure built in 1907 as three row houses. Avenue des Pins is a street that was created in 1907 after the demolition of existing buildings to facilitate access to the Mont-Royal as suggested by Frederick Law Olmsted. Today it is a collection of bad buildings and exposed firewalls that never met the promises of a prestigious boulevard leading to the Mountain.

Theatre is about fugacity, a succession of unique moments that barely survive in the memories of those who were there. Theatres are ghostly figures that have witnessed what we are about to forget. In reconstructing the Théâtre de Quat'sous the designers were specifically asked to incorporate whatever we could find from the original building to help the new one evoke those memories. The designers chose to sample textures, images, colours and materials from a cultural inventory of the theatre and mapped them on the assemblage of required volumes (stage, house, foyer, crossover, control booth and rehearsal). Recycling on site stones, slate, wood, bricks, marble and furniture becomes part of a strategy of cultural sustainability. New materials include silkscreened glass, black brick and perforated aluminum that contribute to make Théâtre de Quat'sous a ghostly figure accumulating memories.

在保罗·比索诺的指导下，一群演员于1963年创立了卡特索斯剧院。他们在蒙特利尔宾斯大道上买下了一个犹太教堂，将它改造成一家小剧院。年轻的创造者一直在此上演着大胆的剧情，其中包括1968年5月上演的神秘的"奥斯帝德克"。

安全问题、使用者的舒适问题和空间需求促进了卡特索斯剧院的拆除和重建。多项研究表明：对建于1907年所建造的砖木结构建筑进行翻修是不可行的。1907年，在景观设计师弗莱德里克·劳·奥姆斯特德的建议下，宾斯大街上的原有建筑被拆除，开通了通往皇家山区的大道。现在大街上遍布破旧的建筑和裸露的防火墙，完全丧失了往日林荫大道的风采。

剧院是不安定的，人们不会记得它所经历的独特变迁。剧院像幽灵一般见证着我们所遗忘的事情。在卡特索斯剧院的重建过程中，设计师被要求尽量保留原有建筑的风韵，让新建筑留存从前的记忆。

设计师从剧院的文化财产清单中选择了材质、图形、色彩和材料样品，将它们装配在必需的空间（如舞台、剧院、大厅、天桥、控制室和排练室）里。现场回收的石材、石板、木材、砖块、大理石和家具成为了文化延续策略的一部分。新材料包括丝网玻璃、青砖和穿孔铝板，它们共同将剧院塑造成了一个富有回忆的幽灵形象。

1. A corner of the façade
2. Sun terrace
3. The frontage of the theatre, a new building full of memories

1. 外立面一角
2. 阳光露台
3. 剧院正面外观，一个富有回忆的新建筑

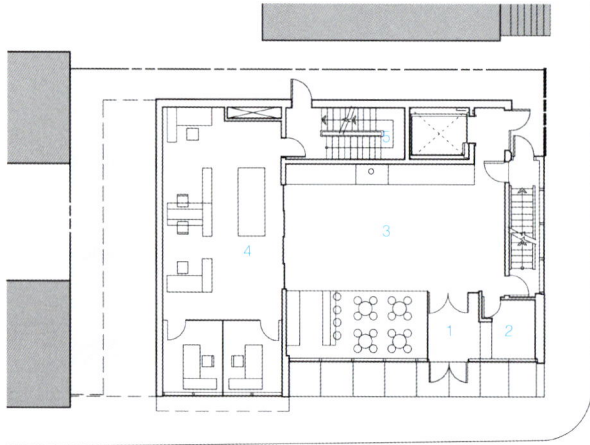

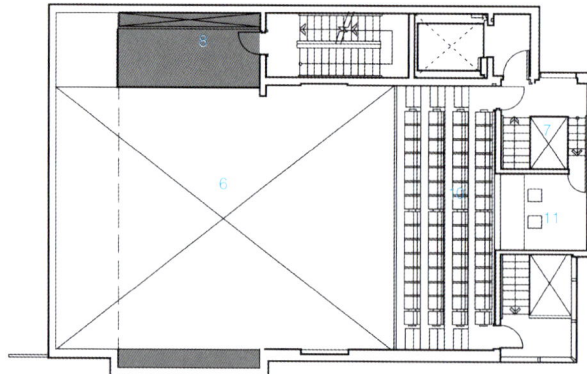

1. Hallway | 1. 门厅
2. Ticket office | 2. 售票处
3. Foyer | 3. 大厅
4. Administration | 4. 行政办公室
5. Emergency stairway | 5. 紧急楼梯
6. Stage | 6. 舞台
7. House | 7. 观众席
8. Offstage | 8. 后台
9. Advertising window | 9. 广告窗
10. Mezzanine | 10. 夹层
11. Control booth | 11. 控制室

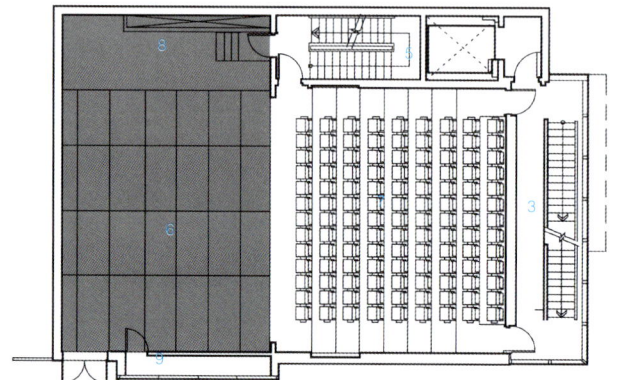

1. In the orange sunlight, the building sits quietly in the corner
2. The staircase and red walls that connect each floor
3. Interior passage are decorated with red, white and blue colours

1. 橘红色的阳光下，剧院静静矗立在街角
2. 连接各层的楼梯与红墙
3. 红白蓝三色映衬的内部通道

4

5

1. Hall
2. Lounge hall
3. The soft lighting in lounge hall
4. Administration office of the theatre
5. Tables and chairs in the lounge hall

1. 大厅
2. 休息厅
3. 休息厅柔和的照明设计
4. 剧院行政管理办公室
5. 休息厅内摆设的桌椅

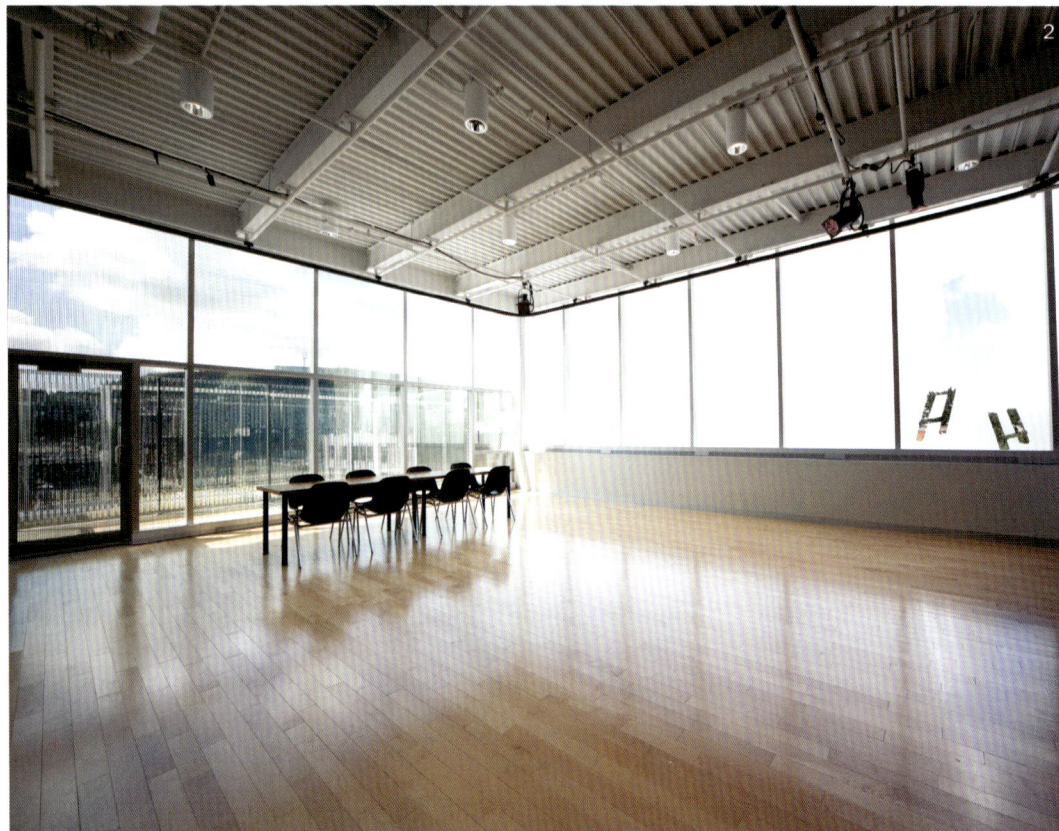

1. The dark-grey walls contrast with the red chairs
2. The rehearsal room has abundant daylight and is bright and spacious
3. Offstage dressing room
4. Offstage changing room
5. Washing stations in the washing room

1. 演出中心暗灰的墙面映衬着鲜红色座椅
2. 排练室自然光线充足，通透宽敞
3. 后台化妆间
4. 后台更衣室
5. 盥洗室洗手台

明洞剧院 # Myeongdong Theatre

Location: Seoul, South Korea **Designer:** Samoo Architects & Engineers **Completion date:** 2009
Photos©: Park Young-chae **Construction area:** 1,286.5 square metres

项目地点：韩国，首尔 设计师：Samoo建筑工程事务所 完成时间：2009年 摄影师：朴勇在 建筑面积：1,286.5平方米

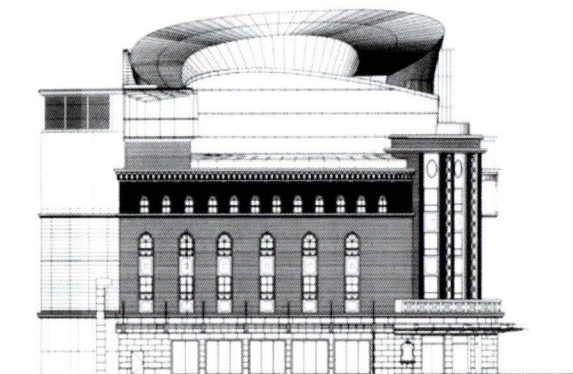

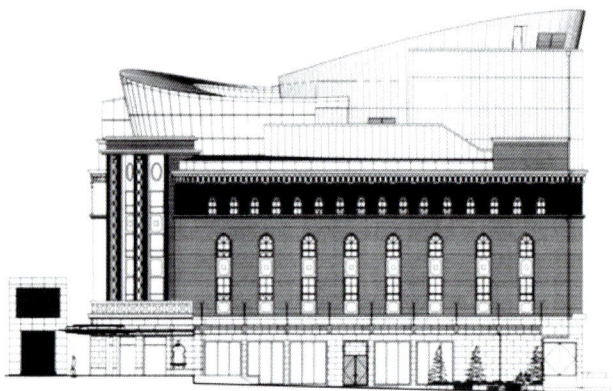

Built in 1930s, the former theatre building had been the cradle of culture and arts before the National Theatre of Korea was relocated from Myeongdong to Jangchungdong in 1973. It was acquired by Daehan Investment & Finance in late 1975 and used as an office until 2003. Due to the soaring land price in Myeongdong and increasing propensity to consume, the former Myeongdong National Theatre building was under threat of being torn down and being replaced by a high-rise commercial building. However, the exterior wall was preserved thanks to the campaign for preserving the building and the purchase by Ministry of Culture, Sports and Tourism. The new cultural interface of the newborn Myeongdong Theatre holds the memories of the past and the vital energy of the district and has become an artistic beacon. The exterior wall, which keeps the memory of Myeongdong, was preserved and a vessel of regeneration (auditorium mass) was created for the culture of Myeongdong streets in the past, present and the future. The energy of the streets was brought inside the building and filled between the wall of the past and the auditorium mass.

For an urban space open to people, the auditorium mass was lifted above a floor to create a large lobby. The natural lighting pouring down through the skylight, which has been installed between the hall and the auditorium mass, runs along down the mass and reaches the lobby, making the narrow space into one enriched by light. As for the auditorium, the balconies of two floors are surrounded in a three-dimensional way and compose a friendly performance hall in the shape of a horseshoe.

The Myeongdong Theatre project is not a simple restoration; its significance lies in the sense that it is the rebirth of the urban, architectural and cultural space for artists, merchants and Seoul citizens who miss Myeongdong as the cultural centre of the past.

在韩国国家剧院于1973从明洞搬迁到奖中洞之前，建于19世纪30年代的剧院大楼一直是文化艺术的摇篮。大韩融资于1975年买下了大楼做办公之用，一直到2003年。由于明洞地价的不断上升和消费倾向的不断增长，前明洞剧院大楼一直处于被拆迁、被高层商业楼替代的危险。但是，多亏了建筑保护运动和文体旅游部的购买，大楼的外墙得以保留。重生的明洞剧院保存着从前的记忆，同时还拥有该地区的生机和活力，成为了一个艺术明灯。饱含明洞历史的外墙得以保留，重建的大礼堂连接了明洞街道的过去、现在和未来。街道的活力被引入了兼职内部，填满了旧墙和大礼堂之间的空隙。

作为一个向公众开放的城市空间，大礼堂提升了一层，形成了一个大厅。自然光透过天窗倾泻下来，射入大厅，让狭窄的空间充满了光线。礼堂两层楼上的包间呈三维造型环绕着表演大厅，形成了一个马蹄铁形。

明洞剧院不仅是一个简单的翻新工程，它的重要性在于城市、建筑、文化空间的重生，以及其对将明洞作为文化中心的艺术家、商人和首尔市民的意义。

1. The theatre is an artistic beacon in the commercial buildings
2. A horseshoe architectural form
3. The exterior of the building – the architecture expresses history and contemporary

1. 剧院是商业楼群中的"艺术明灯"
2. 马蹄形的建筑造型
3. 剧院外观——建筑承载的是历史与现代

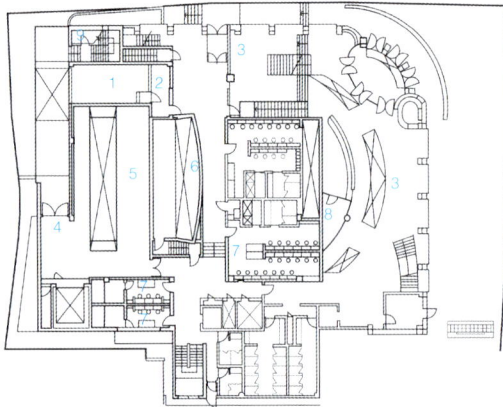

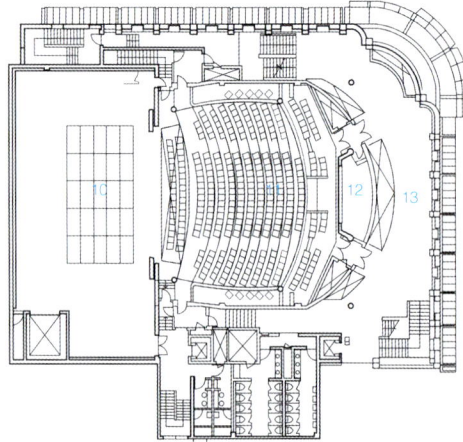

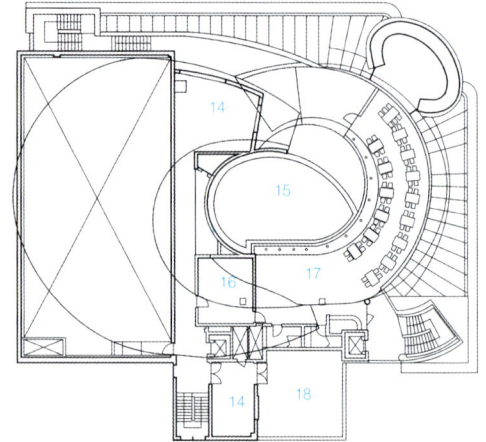

1. Disaster preventing room
2. Guard room
3. Lobby
4. Space for loading and unloading
5. Under stage
6. Orchestra pit
7. Make-up room
8. Office
9. Storage
10. Stage
11. Seat
12. Modulation
13. Hall
14. Air conditioning room
15. Roof garden
16. Kitchen
17. Restaurant
18. Cooling tower

1. 避难室
2. 保安室
3. 大堂
4. 装卸区
5. 舞台下层
6. 乐池
7. 化妆室
8. 办公室
9. 储藏间
10. 舞台
11. 观众坐席
12. 调控室
13. 大厅
14. 空调室
15. 屋顶花园
16. 厨房
17. 餐厅
18. 冷却塔

명동예술극장 개관식 2009년 6월 5일

2

1 Roof garden can also be used as a small performing stage   1. 屋顶花园，也可用作小型演出场地
2. The building sits quietly along the street in the sunshine   2. 阳光下，静静地矗立在街边的建筑

1. On the first floor, the original wall can be seen through the hall's glass wall
2. Ticket office and Information centre on the ground floor
3. Sunlight spills into the hall through the skylight

1. 二层大厅玻璃墙体展示着原建筑的墙体
2. 一层大厅，售票处与信息咨询服务
3. 阳光透过屋顶天窗洒入大厅

1 Seats viewed from the stage
2. Three-level audience seats

1. 演出中心观众席
2. 三层观众坐席

歌词剧院 # Lyric Theatre

Location: Oklahoma City, USA **Designer:** Elliott + Associates Architects **Completion date:** 2007
Photos©: Courtesy of Elliott + Associates Architects **Area:** 1,087 square metres

项目地点：美国，俄克拉何马城 设计师：艾略特建筑事务所 完成时间：2007年 摄影师：艾略特建筑事务所提供 面积：1,087平方米

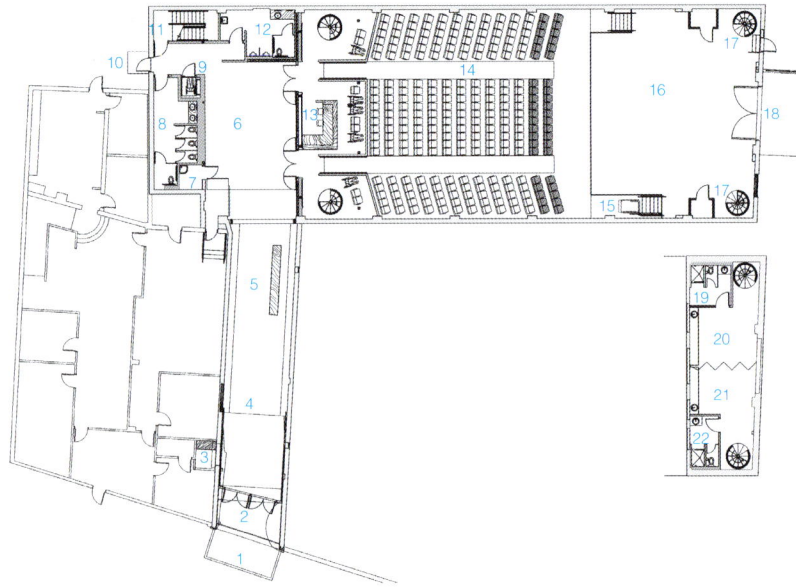

Lyric Theatre is a professional summer stock company founded in 1963 and the only professional musical theatre in Oklahoma. For over 40 years, Lyric Theatre has produced classic and contemporary musicals. There is seating for 278. Total area is 1,087 square metres.

The definition of "lyric" is the word of a song. As an audience member theatre is dramatic, the unexpected, surprising, musical, lyrical, colourful, fantasy, and is about story telling, costumes and grease paint, acting, the spoken word and transporting you to another time and place.

The building had been renovated so many times since 1935 that there was no historic character remaining. Historic preservation was not an option. The designer chose instead to acknowledge the "ghost" of the past. A neon outline "ghost" marquee was designed in the spirit of the original. Exposed brick walls acknowledge the building shell construction and make art of heater cavities and conduit locations once hidden behind long ago destroyed plaster. The original fireplace in the lobby is now a surprise in the men's room. A new wood lath ceiling adds to the raw character of the entrance lobby. An original terrazzo ramp remains and connects outside with inside at the entrance. The concept of "pools" of light creates a fitting theatrical entry experience.

The theatre outer-lobby is designed to have changing light colour that corresponds with the current performance. Plasma screens provide changing donor recognition. The outer lobby is punctuated by eight-inch tall changing LED lighted letters. Toilets remain raw with exposed brick and structure, and use coloured light to continue the theatrical qualities. The theatre space maintains the original riveted steel bow-string trusses and includes the addition of lighting balcony, catwalks and a tension wire grid above the stage.

Two discoveries were made in the basement. One is a newspaper remains exposed in one of the beams. Secondly, a construction worker painted the date 6-5-35 on the concrete wall reminding us of his presence.

1963年，一家专业夏令剧目公司创立了歌词剧院，是俄克拉何马唯一的一家音乐剧院。40多年来，歌词剧院制作了大量古典和现代音乐剧。剧院拥有278个坐席，总面积1,087平方米。

歌词即歌里的词句。对观众来说，剧院充满了戏剧性、未知感、惊喜、音乐、歌词、色彩、幻想、讲故事、表演服、油彩、表演和台词。剧院引领人们走向另一个时间和地点。

自1935年以来，建筑进行了多次翻新，已经丧失了历史特色，因此不需要进行历史保护。设计师通过向"幽灵"致敬来回顾过去。由霓虹灯描绘出的"幽灵"轮廓天幕代表着过去。裸露的砖墙体现了建筑外壳结构，曾破坏石膏墙壁的加热管和导管进行了艺术处理。原来大厅里的壁炉出乎意料地出现在男洗手间。木条天花板为入口大厅增添了原生特色。水磨石坡道得以保留，连接了入口的室内外。"光池"的概念与入口十分相称。

剧院外部大厅的浅色灯光随着表演的剧目而改变。等离子屏幕上显示着捐赠人的信息。外部大厅里装点着不断变化的20厘米高的LED字母灯箱。洗手间裸露的砖石结构和彩色灯光延续了剧院的品质，也具有原生感。剧院空间保持了原有的铆钉铁制弯曲钢架，增添了灯光包厢、天桥和舞台拉力线栅。裸露的砖块标着建筑的外壳。

地下室里有两个特色设计。一是横梁上还有报纸；二是一位工人在混凝土墙上漆上了6-5-35，以时刻提醒我们记住历史。

1. Ghost marquee
2. Main entrance
3. Ticket booth
4. Entrance lobby
5. Concessions
6. Theatre lobby
7. Janitor
8. Women
9. Wheelchair lift
10. Donor garden
11. Stair
12. Men
13. Sound booth
14. Theatre
15. Wheelchair lift
16. Stage
17. Spiral stair
18. Stage entrance
19. Toilet/shower
20. Dressing area
21. Dressing area
22. Toilet/shower

1. 幽灵天幕
2. 主入口
3. 售票处
4. 入口大厅
5. 零售点
6. 剧院大堂
7. 门卫室
8. 女洗手间
9. 轮椅电梯
10. 捐赠人花园
11. 楼梯
12. 男洗手间
13. 音响室
14. 剧院
15. 轮椅电梯
16. 舞台
17. 螺旋楼梯
18. 舞台入口
19. 洗手间/淋浴室
20. 更衣区
21. 更衣区
22. 洗手间/淋浴室

1. Exposed brick walls acknowledge the building shell construction
2. Entrance of the theatre
3. The exterior of the theatre in the dusk

1. 裸露的砖块标着建筑的外壳
2. 剧院入口
3. 黄昏中的剧院外景

2

1, 2. Internal pathways

1、2. 剧院内部通道

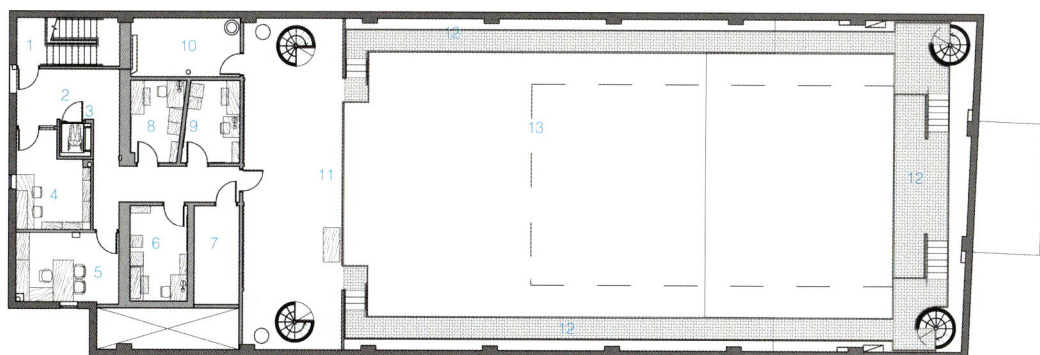

1. Stair	1. 楼梯
2. Corridor	2. 走廊
3. Wheelchair lift	3. 轮椅电梯
4. Library	4. 图书室
5. Office	5. 办公室
6. Voice room 1	6. 语音室1
7. Control room	7. 控制室
8. Voice room 2	8. 语音室2
9. Voice room 3	9. 语音室3
10. Storage	10. 仓库
11. Lighting balcony	11. 灯光包厢
12. Catwalk	12. 天桥
13. Tension wire grid	13. 舞台拉力线栅

1. Make-up room
2. Entrance of rehearsal room
3, 4. LED lighted letters on the wall of the hall
5. Restroom

1. 化妆室
2. 排练室入口
3、4. 大厅墙上的LED字母灯箱
5. 卫生间

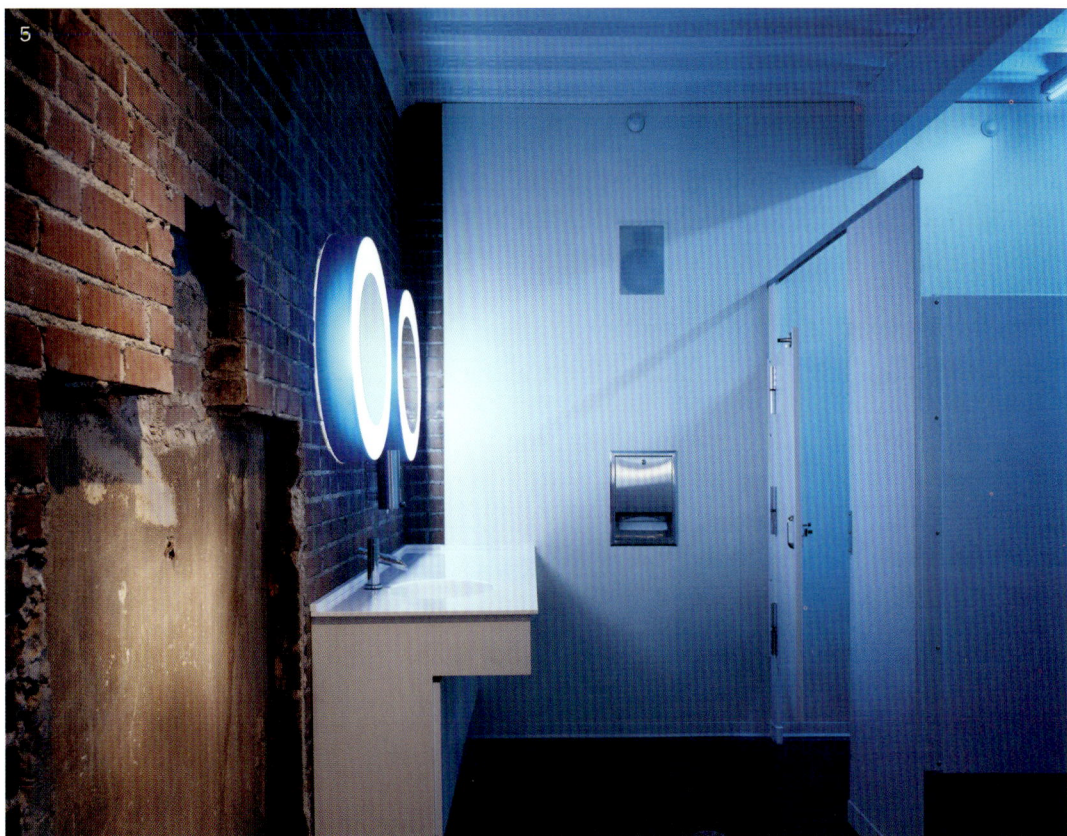

1-4. Performance hall of the theatre

1~4. 剧院演出大厅

赛达德音乐厅 # Cidade Da Música

Location : Rio de Janeiro, Brazil **Designer :** Christian de Portzamparc **Completion :** 2011 **Photos© :** Nelson Kon, Atelier Christian de Portzamparc **Net Area :** 46,000 square metres

项目地点：巴西，里约热内卢 设计师：克里斯汀·德·波特赞姆巴克 完成时间：2011年 摄影师：尼尔森·康 净面积：46,000平方米

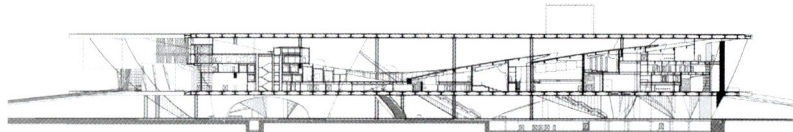

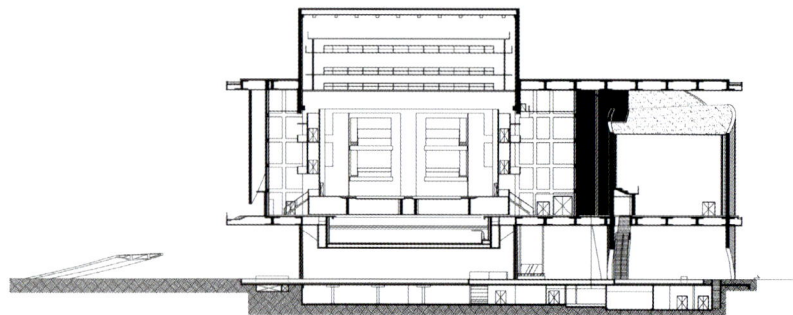

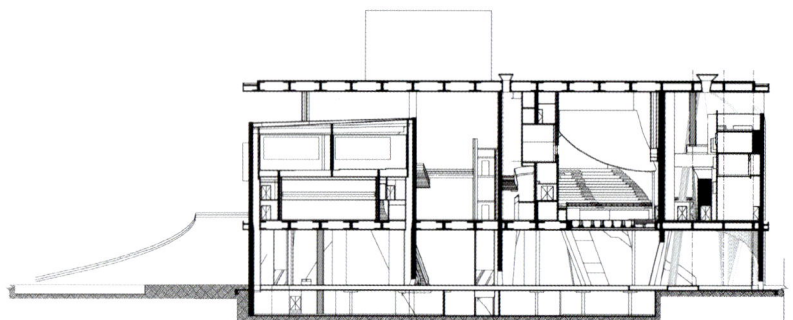

Barra de Tijuca is a new town in Rio de Janeiro's southern outskirts. It is a long plain that lacks strong architectural events and urban marks. The site is structured by two highways that cross. On the centre of this cross imagined by Lucio Costa, the Cidade da Musica will be in the very heart of the new town.

The Cidade da Musica is raised and established on a vast terrace ten metres above a garden designed by Fernando Chacel. This terrace is the public space; it is the gathering place that gives access to all concert rooms, movie theatres, rehearsal rooms, restaurant, library, shops, and the headquarters of the Brazilian Symphonic Orchestra. The Cidade da Musica is seen as a large house on stilts, a great veranda above a garden with ponds, shade and trees. Also, it is homage to an archetype of Brazilian architecture.

Between the two horizontal plates of the roof and the terrace are set the shapes of the concert rooms in an interplay of volumes and voids. The project is an urban signal, a public symbol floating on the plain with a large visibility. The architecture responds also to the beautiful mountain curves of the Siera Atlantica and the line of the sea, this place has the guarantee of being a major landmark of the greater Rio.

Programme includes: Philharmonic Hall of 1,800 places transformable into an Opéra Hall of 1,300 places, Chamber Music Room of 500 places, electroacoustic of 180 places, Headquarters of the Brazilian Symphonic Orchestra, rehearsal rooms, media library, two cinemas with 150 seats, one cinema with 300 seats, restaurant, shops, administration offices, technical spaces and parking lots.

蒂如卡沙洲是里约热内卢南部郊区的新城，狭长的平原缺乏具有特色的建筑和城市地标。建筑场地处在两条高速公路的交叉路口。处在交叉路口中心的赛达德音乐厅是整个新城的中央。

赛达德音乐厅盘踞在一个10米高的巨大平台上，下方是费尔南多·夏赛尔设计的花园。平台作为一个公共空间，通往各个演奏厅、电影院、排练室、餐厅、图书馆、商店和巴西交响乐团的总部。赛达德音乐厅仿佛一座踩着高跷的房子，又像是花园上方的巨大游廊，下方是水池和树荫。它还向巴西建筑原型表达了敬意。

水平屋顶和平台之间是音乐厅的空间。项目是一个醒目的城市信号，是漂浮在平原上的公共标志。建筑与高山的美景、远处的海岸线遥相呼应，是大里约地区的一个主要地标。

项目包括以下设施：1,800坐席的交响乐厅（可转换成1,300坐席的歌剧厅）、500坐席的室内音乐厅、180坐席的电子音乐厅、巴西交响乐团总部、排练室、媒体图书馆、两个150坐席的电影院、一个300坐席的电影院、餐厅、商店、行政办公室、机械室和停车场。

1. Overall view   1. 建筑全景
2. Side view of facade   2. 建筑侧面
3. The building on the vast terrace   3. 建筑处在一个大平台上

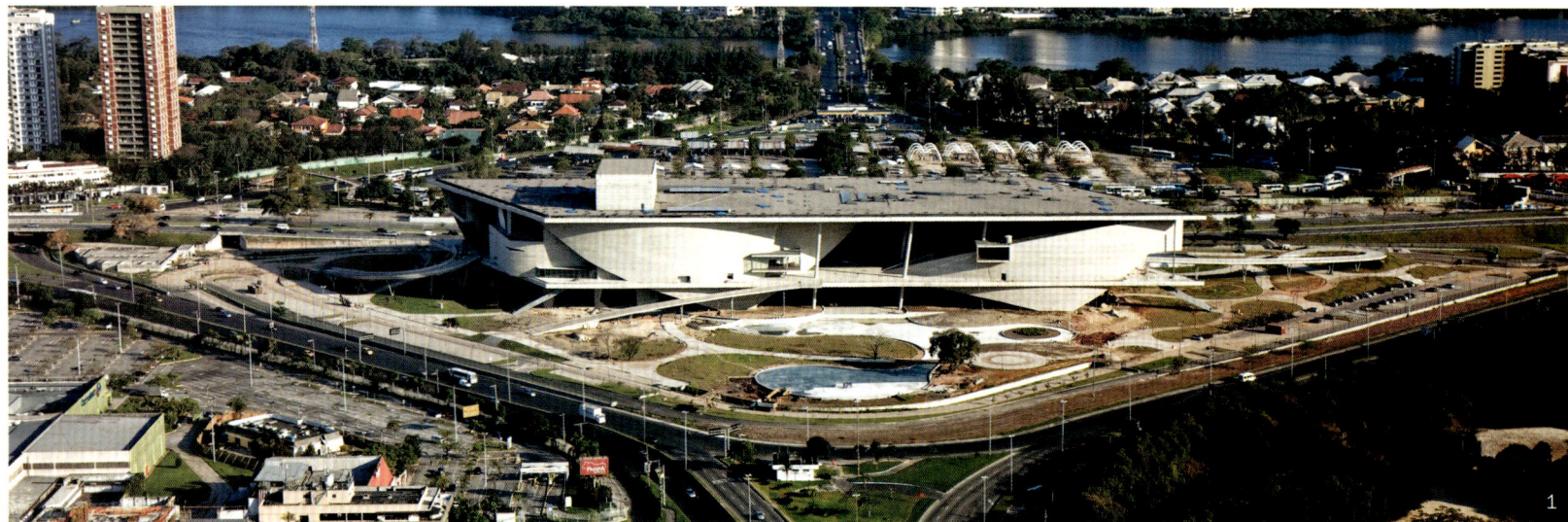

1

2

3

1. Side façade
2. Overall view of façade
3. Exterior detail

1. 建筑侧面
2. 外观全景
3. 外观细部

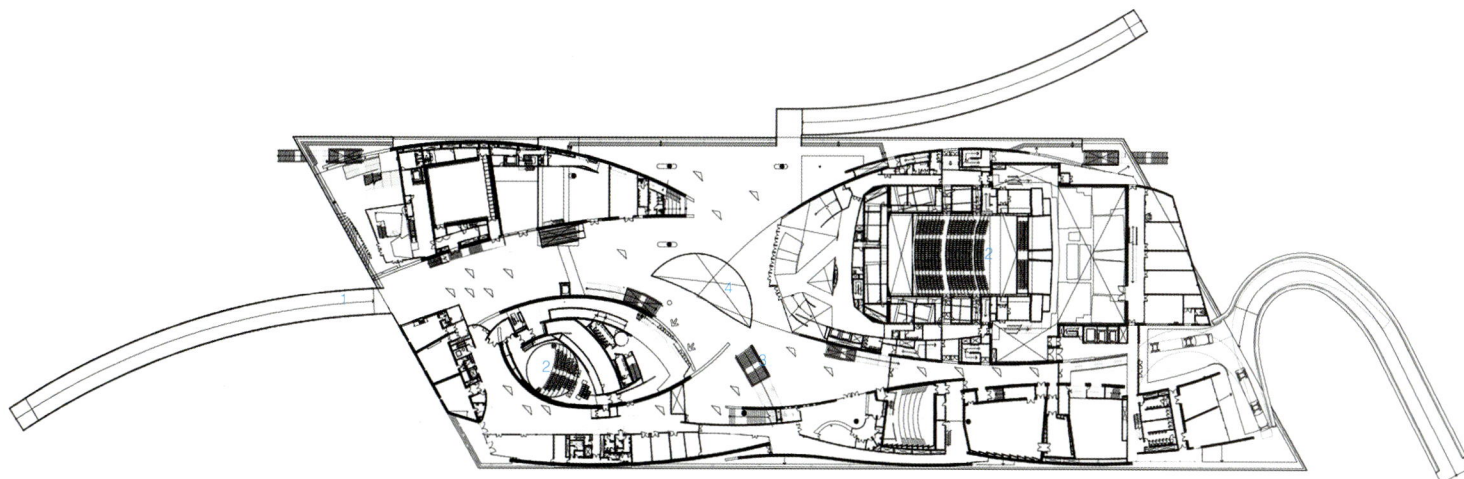

1. Bridge to the first floor
2. Music rooms
3. Stairs
4. Hollow space between different parts of the building

1. 通往二楼的桥梁
2. 音乐室
3. 楼梯
4. 建筑不同部分之间的空隙空间

1. Entrance detail
2. Terrace with open view
3. Space under the terrace

1. 入口细部
2. 开阔的平台
3. 平台下方的空间

A

Agence Nicolas Michelin & Associés (ANMA)
www.anma.fr

ARCHITEKTON
www.architekton.com

Andrej Kalamar, Studio Kalamar
www.kalamar.si

B

Barton Myers Associates, Inc.
www.bartonmyers.com

Bernard Tschumi Architects
www.tschumi.com

C

C. F. Møller Architects
www.cfmoller.com

Christian de Portzamparc
www.chdeportzamparc.com

COBE
www.cobe.dk

Coop Himmelb(l)au
www.coop-himmelblau.at

D

Diamond and Schmitt Architects
www.dsai.ca

Dominique COULON Architects
www.coulon-architecte.fr

Donaire Arquitectos
www.donairearquitectos.com

E

Elliott + Associates Architects
www.e-a-a.com

Enrique Krahe
www.enriquekrahe.com

Építész Stúdió
www.epiteszstudio.hu

F

Foster + Partners
www.fosterandpartners.com

G

gmp – von Gerkan, Marg and Partners Architects
www.gmp-architekten.de

Grimshaw Architects
www.grimshaw-architects.com

H

Henning Larsen Architects and Batteriid Architects
www.henninglarsen.com
www.batteriid.is

Hyunjoon Yoo Architects
www.hyunjoonyoo.com

K

Kuwabara Payne McKenna Blumberg Architects
www.kpmbarchitects.com

L

Les Architectes FABG
www.arch-fabg.com

Lundgaard & Tranberg Arkitekter
www.ltarkitekter.dk

P

Paul de Ruiter
www.paulderuiter.nl

Paul Laurendeau Architecte
www.paullaurendeau.com

Périphérique Architectes
www.peripheriques-architectes.com

R

Rafael Viñoly Architects
www.rvapc.com

REX/OMA
www.rex-ny.com
www.oma.eu

Rios Clementi Hale Studios
www.rchstudios.com

RTKL
www.rtkl.com

S

Saia Barbarese Topouzanov Architectes
www.sbt.qc.ca

Samoo Architects & Engineers
www.samoo.com

Stanton Williams
www.stantonwilliams.com

Stefan Blach/Studio Daniel Libeskind
www.daniel-libeskind.com

U

UN Studio
www.unstudio.com

图书在版编目（CIP）数据

观演建筑 / 殷倩编；常文心译. -- 沈阳 ：辽宁
科学技术出版社，2012.3
    ISBN 978-7-5381-7294-2

    Ⅰ．①观… Ⅱ．①殷… ②常… Ⅲ．①文化建筑—介
绍—中国 Ⅳ．①TU242.2

    中国版本图书馆CIP数据核字(2011)第271733号

出版发行：辽宁科学技术出版社
          （地址：沈阳市和平区十一纬路29号　邮编：110003）
印 刷 者：利丰雅高印刷（深圳）有限公司
经 销 者：各地新华书店
幅面尺寸：240mm×290mm
印　　张：35
插　　页：4
字　　数：50千字
印　　数：1～2000
出版时间：2012年 3 月第 1 版
印刷时间：2012年 3 月第 1 次印刷
责任编辑：陈慈良
封面设计：赵　聪
版式设计：赵　聪
责任校对：周　文
书　　号：ISBN 978-7-5381-7294-2
定　　价：268.00元

联系电话：024-23284360
邮购热线：024-23284502
E-mail: lnkjc@126.com
http://www.lnkj.com.cn
本书网址：www.lnkj.cn/uri.sh/7294